TRANSPORT SIMULATION

ENGINEERING SCIENCES

TRANSPORT SIMULATION
BEYOND TRADITIONAL APPROACHES

Edward Chung and André-Gilles Dumont, Editors

EPFL Press
A Swiss academic publisher distributed by CRC Press

CRC Press
Taylor & Francis Group

Taylor and Francis Group, LLC
6000 Broken Sound Parkway, NW, Suite 300,
Boca Raton, FL 33487

Distribution and Customer Service
orders@crcpress.com

www.crcpress.com

Library of Congress Cataloging-in-Publication Data
A catalog record for this book is available from the Library of Congress.

The authors and publisher express their thanks to the Swiss Federal Institute
of Technology in Lausanne (EPFL) for the generous support towards the
publication of this book.

is an imprint owned by Presses polytechniques et universitaires romandes, a Swiss
academic publishing company whose main purpose is to publish the teaching and
research works of the Ecole polytechnique fédérale de Lausanne (EPFL).

Presses polytechniques et universitaires romandes
EPFL – Centre Midi
Post office box 119
CH-1015 Lausanne, Switzerland
E-Mail: ppur@epfl.ch
Phone: 021 / 693 21 30
Fax: 021 / 693 40 27

www.epflpress.org

© 2009, First edition, EPFL Press
ISBN 978-2-940222-29-2 (EPFL Press)
ISBN 978-1-4200-9509-8 (CRC Press)

Printed in Italy

To Yoko and Yuri

PREFACE

In recent years, the transport simulation of large road networks has become far more rapid and detailed, and many exciting developments in this field have emerged. Transport simulation has broadened from the simulation of automobiles to include pedestrian and rail elements. Advances in transport simulation and the evolving Intelligent Transport System (ITS) are leading to new applications, such as the linking of driving simulators with traffic simulation to give a more realistic environment of driver behaviour surrounding the subject vehicles.

To provide a forum for researchers who are engaged in cutting-edge research in transport simulation would gather from all over the world, the inaugural International Symposium on Transport Simulation (ISTS02) was held in Yokohama, Japan in 2002. There, experts could present their results, discuss research issues and identify future directions of development. Selected papers from ISTS02 were published in *Simulation approaches in transportation analysis – Recent advances and challenges*, edited by R. Kitamura and M. Kuwahara.

Following the success of ISTS02, the second International Symposium on Transport Simulation (ISTS06) was held in Lausanne, Switzerland in 2006. A total of 80 abstracts were received and 43 papers were accepted for presentation, in addition to the four invited speakers. This present book contains a selection of the papers presented at the symposium; each has been reviewed, revised and updated.

This book is divided into four parts. Part I consists of 3 chapters that represent macroscopic, hybrid and vehicle-embedded microscopic traffic simulation. The first paper by Papageorgiou et al. presents the use of a second-order traffic-flow model for multiple traffic operations, such as ramp metering and dynamic traffic assignment. The second paper by Burghout and Koutsopoulos offers a hybrid-modelling framework for loading vehicles at the mesoscopic to microscopic boundary. The new loading method shows superior performance compared to existing approaches. The driving simulator is often used to conduct research on driver behaviour in various traffic conditions which would be otherwise difficult to test in real traffic systems for safety or ethical reasons. The driving simulator offers the benefit of test conditions that are controllable and reproducible; but the vehicle behaviour surrounding the driving simulator must be realistic in order to ensure faithful representation of real driving conditions. The last paper of Part I by Janson Olstam describes the use of a microscopic model to generate and simulate surrounding vehicles for a driving simulator with greater realism.

Part II consists of three chapters that are concerned with lane changing and car following. The first by Ben Akiva et al. summarises a series of advances in lane changing models that include integration of mandatory and discretionary lane changes in a single framework. In Sweden, the 2 + 1 road concept (i.e. two-lane road fitted with an overtaking lane to give safe overtaking opportunities every few kilometres) were found to be a cost-effective way to improve traffic safety. Japanese road administrators have examined the 2 + 1 road concept and an appropriate traffic simulation model for rural 2 + 1 highways is needed. The second chapter by Munehiro et al. describes lane changing behaviour model for rural 2 + 1 highways. This lane changing model was developed based on observed data using video cameras installed on a section of National Highway 38 in Northern Japan. The third chapter by Pottmeier et al. presents a cellular automaton model for multi-lane traffic where vehicles have limited deceleration capability. This extension of the model reduces the probability of crashes. The authors categorise the different behaviour of drivers into optimistic and pessimistic driving. The former controls the behaviour in free flow and the latter governs the driving at high densities.

People movement is the common concern of the first three chapters in Part III. In the first, Asano et al. proposes a framework for a dynamic pedestrian model, taking into account strategic route-choice decisions and multi-directional flow propagation. A modified cell transmission model is adopted to represent the physical and dynamic propagation of pedestrian flows. Kretz and Schreckenberg discuss a floor field and agent-based model (F.A.S.T.) of pedestrian motion in the second chapter. They compared their model performance against a school evacuation drill in a primary school and found the simulation results in good agreement with the empirical data. The third chapter by Hato et al. incorporates pattern matching with detailed time-space resolution into data oriented activity simulation. The authors use dot-matrix and scoring methods for analysing and matching travel-activity patterns. The data-oriented simulation model produces good results when applied to a pedestrian network in Matsuyama City, Japan. The fourth and last paper of Part III is concerned with simulation of urban rail operations. Koutsopoulos and Wang present SimMETRO, a microscopic, dynamic and stochastic simulator of urban rail operations. SimMETRO is specifically designed for service-performance analysis and the evaluation of operations and strategies for real-time control of subway systems. Calibration of such models is critical before their actual use. The authors propose an approach for the calibration of model parameters and inputs, such as dynamic arrival and alighting rates. The results from a case study indicate that the calibration methodology uses the available information effectively to estimate these inputs and parameters and improves the accuracy of the simulation results.

The last part of this book, Part IV, has two chapters. The first by Horiguchi et al. describes the modelling of traffic flow on an expressway toll plaza. As the share of vehicles equipped with on-board Electronic Toll Collection (ETC) units increases, manual toll booths are gradually being replaced by ETC booths in Japan. This increases efficiency, but the interaction between traffic conditions and design of the toll plaza also raises safety concerns (as vehicles have to select the appropriate toll gate). The authors conducted a video survey of vehicle movement in the toll plaza and developed a lane-choice behaviour model in toll plazas which was implemented as an extension module of a microscopic traffic simulation model, AVENUE. Case studies using this simulation model for high ETC penetration scenarios were evaluated and important implications in terms of safety and efficiency were obtained. In the last chapter of this book, Balakrishna et al. present a systematic OD estimation-problem formulation that does not rely on the traditional linear-assignment matrix approximation. Instead, the complex transformations that map OD flows to traffic measurements are captured implicitly through the black-box assignment model itself. They demonstrate the feasibility of the developed methodology through case studies involving the DynaMIT simulation-based traffic assignment model.

This book brings together transport simulation models of different levels (macroscopic, mesoscopic and microscopic) and different modes (automobile, rail and pedestrian). I hope this book will prove to be accessible to engineers and practitioners, and also of interest to the research community. It is hoped that this book will aid in further development and application of transport simulation models in transport planning, management and control.

I would like to thank the many individuals who contributed to reviewing the papers in this book – apologies for not including them all by name. Special thanks go to Ashish Bhaskar, for helping with organisation of the Symposium and assistance during this project. I have had the benefit of working with my mentor, Prof Masao Kuwahara and I thank him for all his support. Finally, I would like to thank all the contributors of this book for making it a reality.

Edward Chung
Lausanne, Switzerland
January 2009

ABOUT THE AUTHORS

Hirokazu Akahane received a Dr. Eng. degree from the University of Tokyo in 1986. He has been professor at the Chiba Institute of Technology since 1995. His major research fields include road traffic control and management. Prof. Akahane is member of JSCE, JSTE and IATSS.
E-mail: akahane@ce.it-chiba.ac.jp

Yasuo Asakuara is professor at the Department of Architecture and Civil Engineering and the Graduate School of Science and Technology of Kobe University. He obtained his bachelor and doctoral degrees from Kyoto University in 1981 and 1988, respectively. He joined Ehime University as a lecturer in 1988, and was promoted to associate professor and then professor, before moving to Kobe University in 2002. Professor Asakura research interests are transport network analysis with a focus on network reliability and route choice modelling, and travel behavioural research with a focus on the behavioural monitoring using mobile communication instruments.
E-mail: asakura@kobe-u.ac.jp

Miho Asano is a researcher at ITS division, National Institute for Land and Infrastructure Management (NILIM), Ministry of Land, Infrastructure, Transport and Tourism (MLIT) in Japan. Her research interests are traffic and pedestrian simulation, Advanced Cruise-assist Highway Systems, traffic signal control, and so on. She received a Dr. Eng. from the University of Tokyo in 2007. Currently she is involved in technical development and deployment of Japanese national project called as Smartway, which are to achieve safe and efficient transport system by using road-infrastructure communication.
E-mail: asano-m92ta@nilim.go.jp

Motoki Asano is the Director for Cold Region Technology Development Coordination, Civil Engineering Research Institute for Cold Region, Public Works Research Institute (PWRI). He holds a Ph.D. degree from Hokkaido University (2005). Until 2007, he was Director of the Traffic Engineering Division of the PWRI.
E-mail: m-asqno@ceri.go.jp

Ramachandran Balakrishna is a Research Transportation Engineer at Caliper Corporation, where he develops and enhances the dynamic traffic and transportation modeling capabilities of TransCAD and TransModeler. He received his undergraduate degree in Civil Engineering from the Indian Institute of Technology, Madras

and earned the M.Sc. and Ph.D. degrees in Transportation Systems from MIT. His research interests span dynamic traffic assignment, traffic simulation, dynamic demand matrix estimation, route choice and demand modeling, and discrete choice analysis. He has also worked extensively on the calibration and validation of traffic simulation models, and his research has appeared in several international journals and conferences.
E-mail: rama@caliper.com

Moshe Ben-Akiva is the Edmund K. Turner Professor of Civil and Environmental Engineering at the Massachusetts Institute of Technology (MIT), and Director of the MIT Intelligent Transportation Systems Program. Prof. Ben-Akiva's research interests include discrete choice and demand modeling, and network modeling and simulation. Two MIT traffic simulators were developed under his supervision: MITSIMLab, and DynaMIT. Prof. Ben-Akiva has published two books, including *Discrete Choice Analysis*, and about 200 papers. He holds a Ph.D. degree in transportation systems from MIT, has received honorary degrees from the Universite Lumiere Lyon, University of the Aegean, and the KTH-Royal Institute of Technology in Sweden, was awarded the Lifetime Achievement Award from the International Association for Travel Behavior Research, and was the recipient of the Jules Dupuit prize from the World Conference on Transport Research Society.
E-mail: mba@mit.edu

Wilco Burghout is Senior Researcher at the Centre for Traffic Research in the School of Architecture and the Built Environment at the Royal Institute of Technology (KTH) in Stockholm. He received his M.Sc. in computer science from the Twente Technical University in the Netherlands and Ph.D. degree in Transportation from KTH, Stockholm. The main theme in his research is simulation-based traffic modeling; microscopic, mesoscopic and hybrid simulation.
E-mail: wilco@ctr.kth.se

Charisma Choudhury is an Assistant Professor in the Department in Civil Engineering at Bangladesh University of Engineering and Technology (BUET) and a Postdoctoral Research Associate at Massachusetts Institute of Technology. She received a Ph.D. in Transportation Systems from the Massachusetts Institute of Technology in 2007. Dr. Choudhury's research is in the areas of behavioral modeling and discrete choice analysis, microsimulation and transportation in developing countries.
E-mail: cfc@mit.edu

Eiji Hato is professor at the Department of Urban Engineering of the University of Tokyo.
E-mail: hato@bin.t.u-tokyo.ac.jp

Ryota Horiguchi is CEO of i-Transport Lab. Co., Ltd. He received a Dr. Eng. degree for the development of traffic simulation model in 1996 from the University of Tokyo. He started his company in 2000 for consultation and system development in

ITS market. His major interest is in traffic simulation, traffic information processing, probe vehicles information system, etc.
E-mail: horiguchi@i-transportlab.jp

Toshio Kamiizumi is an engineer at the Urban & Transportation Department, Kyushu Branch Office, Pacific Consultants, Co., Ltd. Prior to this position, he worked for the Roads & Highways Department at the Transportation Main Office of the same company.
E-mail: toshio.kamiizumi@tk.pacific.co.jp

Masuo Kashiwadani, Dr. Eng., is Professor of the Department of Civil and Environmental Engineering of **Ehime** University (Japan).
E-mail: kashiwa1@eng.ehime-u.ac.jp

Haris N. Koutsopoulos is Professor and Head of the Traffic and Logistics Division in the School of Architecture and the Built Environment at the Royal Institute of Technology (KTH) in Sweden. He received his Sc.Eng. from the National Technical University of Athens, Greece, and the M.Sc. and Ph.D. degrees in Transportation Systems from MIT. His research focuses on ITS, simulation-based dynamic traffic assignment, traffic simulation, simulation of urban rail operations, modeling driver behavior, and calibration of simulation models. He has published extensively in these areas and he is a member of Traffic Flow Theory Committee at TRB, and in the editorial board of two international journals.
E-mail: hnk@infra.kth.se

Tobias Kretz studied physics at Karlsruhe University, where he graduated in 2003 with a thesis in particle theory. He then joined the group of Prof. Schreckenberg at Duisburg-Essen University where in 2007 he received his Ph.D. with the thesis "Pedestrian Traffic – Simulation and Experiments". Since July 2007 he works as product manager for the simulation of pedestrians in the microscopic traffic simulation VISSIM for PTV AG in Karlsruhe.
E-mail: Tobias.Kretz@PTV.de

Masao Kuwahara is a professor at Institute of Industrial Science, University of Tokyo. After receiving Ph.D. from University of California, Berkeley, he has over 20-year experience in research and education on traffic engineering. His major interests are dynamic network analysis, traffic simulation, highway capacity, traffic signal control, ITS, etc. He has been appointed a member of various committees of ministries, local governments, and public corporations on transportation planning and traffic management. He has also served as an International Advisory Committee member of ISTTT as well as an editor in several international journal including ITS Journal, Transportmetrica, and Sustainable Transportation.
E-mail: kuwahara@iis.u-tokyo.ac.jp

Kazunori Munehiro is Senior Researcher of the Traffic Engineering Research Team at the Civil Engineering Research Institute for Cold Region, Public Works Research

Institute (PWRI). In the recent past, he has worked for the Traffic Engineering Division of the Civil Engineering Research Institute of Hokkaido.
E-mail: k-munehiro@ceri.go.jp

Johan Janson Olstam received his Master of Science degree and Licentiate of Engineering degree from Linköping University, Norrköping, Sweden, 2002 and 2005, respectively. His research focuses on traffic simulation and especially simulation of surrounding vehicles in driving simulators, which was the topic for his licentiate thesis and is the topic for the forthcoming Ph.D. thesis. He also works and teaches within other areas in the traffic engineering field, for example volume-delay functions and traffic signal control. He is currently working at Linköping University and the Swedish National Road and Transport Research Institute (VTI).
E-mail: johan.janson.olstam@vti.se

Markos Papageorgiou has been a Professor at the Technical University of Crete, Chania, Greece since 1994. From 1988 to 1994 he was a Professor of Automation at the Technical University of Munich. He was a Visiting Professor at the Politecnico di Milano, Italy (1982), at the Ecole Nationale des Ponts et Chaussées, Paris (1985-1987), and at MIT, Cambridge (1997, 2000); and a Visiting Scholar at the University of California, Berkeley (1993, 1997, 2000). Prof. Papageorgiou is the Editor-in-Chief of *Transportation Research – Part C* and an Associate Editor of *IEEE Control Systems Society – Conference Editorial Board.* He was Chairman (1999-2005) of the IFAC Technical Committee on Transportation Systems. He is a Fellow of IEEE. He received a DAAD scholarship (1971-1976), the 1983 Eugen-Hartmann award from the Union of German Engineers (VDI), and a Fulbright Lecturing/Research Award (1997).
E-mail: markos@dssl.tuc.gr

Ioannis Papamichail received the Dipl.-Eng. (honors) degree in Chemical Engineering from the National Technical University of Athens, in 1998 and the MSc in Process Systems Engineering with distinction and Ph.D. in Chemical Engineering from the Imperial College London in 1999 and 2002 respectively. From 1999 to 2002, he was a Research and Teaching Assistant at the Centre for Process Systems Engineering, Imperial College London. From 2004 to 2005, he was an Adjunct Lecturer, and since 2005 he has been a Lecturer, at the Technical University of Crete, Chania, Greece. Dr Papamichail received an Eugenidi Foundation scholarship (1998).
E-mail: ipapa@dssl.tuc.gr

Andreas Pottmeier is employed at the TraffGo Road GmbH as a consultant and developer. Between 2000 and 2007 he worked at the Physics of Transport and Traffic Group at University Duisburg-Essen as a student assistant and postgraduate. He obtained his Ph.D. in 2007 in the field of theoretical physics. His research interests are in transport problems especially of vehicular traffic and prognosis as well as geospatial content and GIS-services.
E-mail: pottmeier@traffgoroad.com

Mamoru Sasaki is an engineer for the Urban and Transportation Department, Hokkaido Branch Office, Pacific Consultants, Co., Ltd.
E-mail: mamoru.sasaki@ss.pacific.co.jp

Andreas Schadschneider is professor for theoretical physics at Cologne University. He has obtained his Ph.D. in 1991 in the field of solid state physics. Since more than 15 years he is working on problems of non-equilibrium physics. Here his focus are transport problems with interdisciplinary applications, e.g. in traffic engineering, biology and social dynamics. He has organized several international conferences and is author in various review articles on the application of methods from physics to traffic and transport problems.
E-mail: as@thp.uni-koeln.de

Michael Schreckenberg, born 1956 in Düsseldorf, studied theoretical physics at the University of Cologne, where he received his Ph.D. in statistical physics in 1985. In 1994 he moved to the University Duisburg-Essen, where he became the first German professor for Physics of Transport and Traffic in 1997. Since more than ten years he is working on modelling, simulation and optimisation of large-scale transportation systems, especially road traffic and the influence of human behaviour on it. His current activities include online traffic-forecasts on the highway network in North Rhine-Westphalia, the reaction of drivers on traffic information and the analysis of a crowd of panicking people.
E-mail: schreckenberg@ptt.uni-due.de

Takahiro Shitama is a Researcher in Traffic Engineering Division of Expressway Research Institute. He received a M. Eng. degree from Chuo University in 1997.
E-mail: t.shitama.aa@ri-nexco.co.jp

Agachai Sumalee (www.cse.poly.edu.hk/~ceasumal) is an Assistant Professor at Department of Civil and Structural Engineering, The Hong Kong Polytechnic University. His research interests include transport network modeling, network optimization, network reliability analysis, road pricing, and sustainable transport. Agachai secured research grants from various sources including the UK DfT, EPSRC, Volvo Research Foundation, European Commission, and Hong Kong Research Grant Council. Agachai currently serves as a member of Mass Transit System sub-committee of Thailand State Railway Authority and also a committee of network modeling committee of the US Transportation Research Board. He is currently an Associate Editor of *Networks and Spatial Economics*.
E-mail: ceasumal@polyu.edu.hk

Christian Thiemann is a pre-doctoral fellow in the Department of Engineering Sciences and Applied Mathematics at Northwestern University, USA. His research interests are in human transportation networks and the spatial dynamics of infectious diseases. He received his diploma (M.Sc. equivalent) degree in physics from the Georg-August-University Göttingen, Germany in 2008 and worked as a student assistant at the Physics of Transport and Traffic Group at University Duisburg-Essen,

Germany in 2005/06, and the Chair for Traffic Modelling and Econometrics at the Dresden University of Technology, Germany in 2006/07.
E-mail: thiemann@northwestern.edu

Tomer Toledo is a Senior Lecturer at the Transportation Research Institute and Faculty of Civil and Environmental Engineering at the Technion – Israel Institute of Technology in Haifa, Israel. He received a Ph.D. in Transportation Systems from the Massachusetts Institute of Technology in 2002. Dr. Toledo's research is in the areas of large-scale traffic simulation, driver behavior and safety, and intelligent transportation systems.
E-mail: Toledo@technion.ac.il

Toshiya Uzuka is Deputy Director of the Road First Division, Otaru Construction Development Department, Hokkaido Development Bureau, Ministry of Land, Infrastructure, Transport and Tourism (MLIT). He was recently Chief of Planning First Section, Road Planning Division, of the MLIT.
E-mail: uzuka-t22aa@hdk.mlit.go.jp

Yibing Wang received a Ph.D. degree in Control Theory and Applications from Tsinghua University, China. He was with the Dynamic Systems and Simulation Laboratory, Technical University of Crete, Greece, where he was a Postdoctoral Researcher from 1999 to 2001 and a Senior Research Fellow from 2001 to 2007. He is currently a Senior Lecturer with the Department of Civil Engineering, Monash University, Melbourne, Australia. He has published more than 20 international journal papers and book chapters on road traffic modelling, surveillance and control. Dr. Wang is an Associate Editor for the IEEE *Transactions on Intelligent Transportation Systems*, and the Book Review Editor of *Transportation Research Part C: Emerging Technologies*.
E-mail: yibing.wang@eng.monash.edu.au

Zhigao Wang received his Ph.D. in Transportation from Northeastern University in Boston, MA in 2006. He subsequently worked as an analyst for Transystems in the Boston area.

Jian Xing is a director of Traffic Research Division of Highway Engineering Research Department in Expressway Technology Center. He graduated from Tsinghua University in China in 1987 and received a Dr. Eng. Degree from the University of Tokyo in 1992. His current projects and research interests are related to motorway traffic operation and management, application of ITS to mitigate traffic congestion and traffic safety on intercity motorways.
E-mail: jian-x@extec.or.jp

TABLE OF CONTENTS

*Ramachandran Balakrishna, Moshe Ben-Akiva,
Haris N. Koutsopoulos*

PART I

Macroscopic to Vehicle-Embedded Microscopic Simulation

THE ROLE OF MACROSCOPIC MODELING IN THE SIMULATION, SURVEILLANCE AND CONTROL OF MOTORWAY NETWORK TRAFFIC

Markos Papageorgiou, Ioannis Papamichail, Yibing Wang

Dynamic and macroscopic discretized analytic state-space models of motorway network traffic are more than mere simulation tools; they can be used as a valuable basis for the employment of powerful methods for surveillance and control, such as a Kalman Filter, various Automatic Control methodologies and gradient-based optimization. In the motorway traffic context, this leads to a transparent, convenient and efficient handling of various significant tasks, including dynamic traffic assignment, traffic state estimation and prediction, real-time parameter estimation, incident alarm, travel-time and congestion estimation and prediction, optimal co-ordinated ramp metering, system-optimal or user-optimal route guidance and integrated traffic control.

1.1 INTRODUCTION

Mathematical modeling is the imitation of the relevant aspects of a process (e.g., traffic flow) by use of appropriate mathematical equations and further logical relationships. When fed with sufficient initial and boundary conditions as well as control inputs and further exogenous variables, a *dynamic* model may produce the evolution of the process state over time. If the modeling equations are appropriately implemented in a computer, the resulting *simulator* can be employed as a cost-effective and convenient tool for multiple uses, such as, in the case of traffic flow, planning of new extended transportation infrastructure; testing the efficiency of various traffic control measures, strategies and systems; comparison of alternatives etc.

Two basic modeling approaches have been pursued in the area of traffic flow; *microscopic* modeling, which describes the longitudinal (car-following) and lateral (lane-changing) movement of individual vehicles; and *macroscopic* modeling, which addresses traffic as a particular fluid with aggregate variables (density, mean speed, flow). Both approaches can be contrasted with respect to a number of aspects:

- *Simulation*: The traffic flow simulation market is dominated by the microscopic approach, possibly due to its direct similarity with the perceived real process as compared to the abstract, mathematically more challenging macroscopic description.

- *Computational effort* is far lower in macroscopic approaches.

- Beyond simulation, macroscopic models of whole traffic networks can be expressed in an *analytic form* that opens the way for the use of a multitude of powerful available surveillance and control methods such as a Kalman Filter, Automatic Control methods, gradient-based optimization etc.

- *Accuracy* depends on the employed validation procedures that are easier to apply in the low-resolution macroscopic models.

- Specific considerations (e.g. impact of various messages to the driver, mixed automated/non-automated traffic flow, impact of infrastructure layout changes etc.) may be more easily and directly incorporated in microscopic models.

Over the decades, various dynamic macroscopic models, mostly in the form of Partial Differential Equations (PDE), of related research efforts have been proposed by Hoogendoorn and Bovy (2001). As the conservation equation is the only exact relationship in traffic flow modeling, it is included in all approaches. In addition, first-order models involve a static speed-density relationship while second-order models address the mean speed dynamics with potentially more realism. Although the technical literature on macroscopic traffic flow modeling is vast and increasing in an accelerated pace, rigorous model validation exercises using real traffic data are surprisingly sparse. Given the largely empirical character of the proposed models, the lack of validation efforts is a shortcoming that cannot be sufficiently emphasized.

Another issue connected to macroscopic models is the space-time discretization of the related PDE in order to enable their numerical solution in digital computers. In numerous cases, sophisticated numerical schemes are employed for a reliable and accurate numerical approximation of the PDE. These approaches, however, typically result in complex computational schemes that require a high computational effort and, moreover, do not lead to analytical discretized models; in other words, these approaches employ a significant effort to approximate the PDE that are all but exact. An alternative, more practicable, approach is to discretize the original empirical PDE by use of simple (e.g., Euler) schemes leading to analytical state-space models that can then be readily validated; implemented in a computer with low computational effort; used as a valuable basis for the analytical derivation of various surveillance and control tools. The main disadvantage of these approaches resides in the fact that any theoretical investigations and results obtained for the PDE are not directly transferable to the discretized model.

Mathematical models are enabling tools for a number of tasks, ultimately providing a more efficient planning and operation of traffic systems. This chapter presents a simple, discretized second-order traffic flow model that has been validated with good results at several instances and has provided a valuable basis for multiple significant operation-related tasks: dynamic traffic assignment, route guidance, surveillance, ramp metering, integrated traffic control etc.

1.2 MACROSCOPIC MODELING OF MOTORWAY NETWORK TRAFFIC

A second-order macroscopic discretized traffic flow model was used for the description of traffic flow on a motorway link. This model is suitable for free-flow, critical and congested traffic conditions and is included in the generic simulation tool METANET (Messmer and Papageorgiou, 1990). It has two distinct modes of operation. When traffic assignment aspects are not considered, it operates in the non-destination-oriented mode. When, on the other hand, traffic assignment is an issue, it operates in the destination-oriented mode.

The motorway network is represented by a directed graph of which the links represent motorway stretches with uniform characteristics, i.e., without on-/off-ramps or major changes in geometry. The nodes of the graph are placed at locations where a significant change in road geometry occurs, as well as at junctions, on-ramps and off-ramps.

The macroscopic description of traffic flow implies the definition of adequate variables expressing the aggregate behavior of traffic at certain times and locations. The time and space arguments are discretized, and the discrete time step is denoted T (typically $T \simeq 10$ s). A motorway link m is divided into N_m segments of equal length L_m (typically $L_m \simeq 500$ m) (Fig. 1.1), such that the numerical stability condition $L_m \geq T v_{f,m}$ holds, where $v_{f,m}$ is the free speed on link m. Each segment i of link m at discrete time $t = kT$, $k = 0,1,...,K$, where K is the time horizon, is macroscopically characterized by the following variables:

- *Traffic density* $\rho_{m,i}(k)$ (veh/lane/km) is the number of vehicles in segment i of link m at time $t = kT$ divided by L_m and by the number of lanes λ_m.

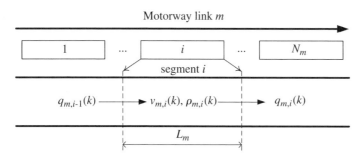

Fig. 1.1 Discretized motorway link.

- *Mean speed* $v_{m,i}(k)$ (km/h) is the average speed of the vehicles included in segment i of link m at time $t = kT$.

- *Traffic volume* or *flow* $q_{m,i}(k)$ (veh/h) is the number of vehicles leaving segment i of link m during the time period $[kT, (k + 1)T]$, divided by T.

Furthermore, for the destination-oriented mode of operation, the following variables are introduced:

- *Partial density* $\rho_{m,i,j}(k)$ is the density of vehicles in segment i of link m at time $t = kT$ destined to reach destination $j \in J_m$, where J_m is the set of destinations reachable via link m.

- Composition rate $\gamma_{m,i,j}(k)$, $0 \le \gamma_{m,i,j}(k) \le 1$, is the portion of traffic volume $q_{m,i}(k)$ or traffic density $\rho_{m,i}(k)$ destined to reach destination $j \in J_m$.

1.2.1 The non-destination-oriented model

In the non-destination-oriented model, the previously defined traffic variables are calculated for each segment i of link m at each time step k by the following equations:

$$\rho_{m,i}(k+1) = \rho_{m,i}(k) + \frac{T}{L_m \lambda_m}\left[q_{m,i-1}(k) - q_{m,i}(k)\right] \tag{1.1}$$

$$q_{m,i}(k) = \rho_{m,i}(k)v_{m,i}(k)\lambda_m \tag{1.2}$$

$$v_{m,i}(k+1) = v_{m,i}(k) + \frac{T}{\tau}\left\{V\left[\rho_{m,i}(k)\right] - v_{m,i}(k)\right\}$$
$$+ \frac{T}{L_m}\left[v_{m,i-1}(k) - v_{m,i}(k)\right]v_{m,i}(k) - \frac{\nu T}{\tau L_m}\frac{\rho_{m,i+1}(k) - \rho_{m,i}(k)}{\rho_{m,i}(k) + \kappa} \tag{1.3}$$

$$V\left[\rho_{m,i}(k)\right] = v_{f,m}\exp\left[-\frac{1}{\alpha_m}\left(\frac{\rho_{m,i}(k)}{\rho_{cr,m}}\right)^{\alpha_m}\right] \tag{1.4}$$

where $\rho_{cr,m}$ denotes the critical density per lane of link m, and α_m is a parameter of the fundamental diagram (Eq. (1.4)) of link m expressing a nonlinear relationship between the mean speed and the traffic density. Furthermore, the time constant τ, the anticipation constant ν, and κ, are further parameters that are equal for all the network links. The parameter values can be determined via a validation procedure described in Section 1.2.4.

This second-order model was essentially proposed by Payne (1971). Equation (1.1) expresses the vehicle-conservation principle, while Eq. (1.2) is the flow equation that results directly from the non-discretized definition of the traffic variables. Equation (1.3) is an empirical dynamic speed equation that describes the dynamic evolution of the mean speed of each segment as an independent variable. Equation (1.4) is also an empirical static relationship between speed and traffic density. Two additional terms can be included in Eq. (1.3) for the modeling of lane drops and

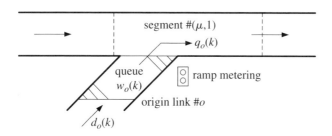

Fig. 1.2 The origin-link queue model.

merging phenomena at on-ramps as suggested by Papageorgiou et al. (1990), and these include the two constant model parameters δ and ϕ. Additionally, the mean speed resulting from Eq. (1.3) is limited from below by the minimum speed $v_{\min}$ in order to avoid unrealistically low flows during congestion.

For origin links, i.e., links that receive the traffic demand d_o and forward it into the motorway network, a simple queue model is used (Fig. 1.2). The outflow q_o of an origin link o depends on the traffic conditions of the corresponding mainstream segment $(\mu,1)$ and the existence of ramp metering control measures. If ramp metering is applied, the outflow $q_o(k)$ leaving origin o during period k, is a portion $r_o(k)$ of the maximum outflow $\hat{q}_o(k)$ that would leave in absence of ramp metering. Thus, $r_o(k) \in [r_{\min,o},1]$ is the metering rate for the origin link o, i.e., a control variable, where $r_{\min,o}$ is a minimum admissible value. If $r_o(k) = 1$, no ramp metering is applied, otherwise $r_o(k) < 1$. The queuing model is described by the following conservation equation:

$$w_o(k+1) = w_o(k) + T[d_o(k) - q_o(k)] \tag{1.5}$$

where $w_o(k)$ (veh) is the queue length in origin o at time kT and $d_o(k)$ (veh/h) is the demand flow at o. The outflow q_o is determined as:

$$q_o(k) = r_o \hat{q}_o(k) \tag{1.6}$$

with

$$\hat{q}_o(k) = \min\{\hat{q}_{o,1}(k), \hat{q}_{o,2}(k)\} \tag{1.7}$$

and

$$\hat{q}_{o,1}(k) = d_o(k) + \frac{w_o(k)}{T} \tag{1.8}$$

$$\hat{q}_{o,2}(k) = Q_o \min\left\{1, \frac{\rho_{\max} - \rho_{\mu,1}(k)}{\rho_{\max} - \rho_{cr,\mu}}\right\} \tag{1.9}$$

where Q_o (veh/h) is the capacity of the on-ramp, i.e., its maximum possible outflow under free-flow traffic conditions in the mainstream, and $\rho_{\max}$ (veh/lane/km) is the maximum density in the network. Thus, the maximum outflow $\hat{q}_o(k)$ is determined by the current origin demand if $\hat{q}_{o,1}(k) < \hat{q}_{o,2}(k)$, or by the geometrical capacity Q_o if the mainstream density is subcritical, i.e., $\rho_{\mu,1}(k) < \rho_{cr,\mu}$, or by the reduced capacity due to congestion of the mainstream if $\rho_{\mu,1}(k) > \rho_{cr,\mu}$.

For a number of reasons, including the modeling of motorway-to-motorway control measures, a similar simple queue model may also be used for certain internal network links, called store-and-forward links. Traffic conditions in destination links are influenced by the downstream traffic conditions that can be provided as boundary conditions for the whole time horizon. Dummy links are auxiliary links of zero length. They do not affect traffic dynamics but are used to decompose complex network topologies or to represent very short motorway connections.

Motorway bifurcations and junctions (including on- and off-ramps) are represented by nodes. Traffic enters a node n through a number of input links and is distributed to the output links according to the following equations:

$$Q_n(k) = \sum_{\mu \in I_n} q_{\mu,N_\mu}(k) \tag{1.10}$$

$$q_{m,0}(k) = \beta_n^m(k) Q_n(k) \forall m \in O_n \tag{1.11}$$

where I_n is the set of links entering node n, O_n is the set of links leaving n, $Q_n(k)$ is the total traffic volume entering n at period k, $q_{m,0}(k)$ is the traffic volume leaving n via outlink m and $\beta_n^m(k)$ is the portion of $Q_n(k)$ leaving n through link m. Thus, $\beta_n^m(k)$ are the turning rates of node n.

If a node n has more than one leaving link, the upstream influence of density has to be taken into account in the last segment of the incoming link (see Eq. (3) for $i = N_m$). This is provided via

$$\rho_{m,N_m+1}(k) = \frac{\displaystyle\sum_{\mu \in O_n} \rho_{\mu,1}^2(k)}{\displaystyle\sum_{\mu \in O_n} \rho_{\mu,1}(k)} \tag{1.12}$$

where ρ_{m,N_m+1} is the virtual density downstream of the entering link m to be used in Eq. (1.3) for $i = N_m$, and $\rho_{\mu,1}(k)$ is the density of the first segment of the leaving link μ. The quadratic form is used to account for the fact that traffic flow of one congested link may spill back into the entering link even if there is free flow in the other leaving links.

If a node n has more than one entering links, the downstream influence of the speed has to be considered according to Eq. (1.3) for $i = 1$. The mean speed value is calculated from

$$v_{m,0} = \frac{\displaystyle\sum_{\mu \in I_n} v_{\mu,N_\mu}(k) q_{\mu,N_\mu}(k)}{\displaystyle\sum_{\mu \in I_n} q_{\mu,N_\mu}(k)} \tag{1.13}$$

where $v_{m,0}$ is the virtual speed upstream of the leaving link m which is needed in Eq. (1.3) for $i = 1$.

1.2.2 The destination-oriented model

When traffic assignment is necessary, the notion of turning rates β_n^m is generalized to the notion of splitting rates. Here, $Q_{n,j}(k)$ is the total traffic volume entering motorway node n at period k destined to reach j. In this situation, the splitting rate $\beta_{n,j}^m(k)$ is the portion of the traffic volume $Q_{n,j}(k)$ that leaves node n at period k through link $m \in O_n$, hence $0 \le \beta_{n,j}^m(k) \le 1$. Thus, for any network node:

$$Q_{n,j}(k) = \sum_{\mu \in I_n} q_{\mu,N_\mu}(k) \gamma_{\mu,N_\mu,j}(k) \quad \forall (n,j) \tag{1.14}$$

$$q_{m,0}(k) = \sum_{j \in J_m} Q_{n,j}(k) \beta_{n,j}^m(k) \quad \forall m \in O_n \tag{1.15}$$

$$\gamma_{m,0,j}(k) = \beta_{n,j}^m(k) Q_{n,j}(k)/q_{m,0}(k) \quad \forall m \in O_n, \forall j \in J_m. \tag{1.16}$$

Note that $\sum_{m \in O_n} \beta_{n,j}^m(k) = 1$ holds, thereby leading to a reduction by one of the number of independent splitting rates at each bifurcation node.

The partial densities in each link are calculated from conservation considerations

$$\rho_{m,i,j}(k+1) = \rho_{m,i,j}(k)$$
$$+ \frac{T}{L_m \lambda_m} \left[\gamma_{m,i-1,j}(k) q_{m,i-1}(k) - \gamma_{m,i,j}(k) q_{m,i}(k) \right] \quad \forall j \in J_m \tag{1.17}$$

while $\gamma_{m,i,j}(k) = \rho_{m,i,j}(k)/\rho_{m,i}(k)$. Similarly, for the queues of origin as well as store-and-forward links, the notion of partial queues $w_{o,j}$ is introduced. Partial queues evolve according to the relationship

$$w_{o,j}(k+1) = w_{o,j}(k) + T \left[\theta_{o,j}(k) d_o(k) - \gamma_{o,j}(k) q_o(k) \right] \tag{1.18}$$

where $\theta_{o,j}(k)$ is the portion of $d_o(k)$ destined to reach j during period k, $\gamma_{o,j} = w_{o,j}(k)/w_o(k)$ and $w_{o,j}(k)$ is the number of vehicles in the queue of origin link o with destination j.

In the case where route guidance takes place at a node n with respect to destination j (using variable message sings or other means), a direction is recommended to the drivers towards this destination (see Sections 1.2 and 1.5.3). Depending on the drivers' compliance rate, this recommendation may affect their behavior. Since the routing message refers to particular destinations, the influence on the route choice is projected to the corresponding splitting rates of the node. At bifurcation node n for destination j, the following splitting rates are defined: $\beta_{N,n,j}^m(k)$ is the nominal splitting of drivers in the absence of any guidance; $\beta_{G,n,j}^m(k)$ is the splitting order by the system, i.e., a control variable; $\beta_{n,j}^m(k)$ is the real splitting according to the drivers' compliance. The relation between these three quantities is modeled by the following equation:

$$\beta_{n,j}^m = (1 - \varepsilon) \beta_{N,n,j}^m + \varepsilon \beta_{G,n,j}^m \tag{1.19}$$

where ε is the compliance rate to the guidance instructions ($0 \le \varepsilon \le 1$).

The described approach uses splitting rates (i.e., turning rates by destination) for traffic assignment purposes, while the use of paths connecting each origin-destination pair in the network is a more popular approach. The advantages of splitting rates over paths include:

- the number of splitting rates in large networks being several orders of magnitude smaller than the number of paths with corresponding implications for simplicity and computation times;

- a splitting rate $\beta_{n,j}^m(k)$ having a direct physical interpretation, as it corresponds to the indications of a VMS (variable message sign) located at network node n and guiding drivers bound for destination j.

Clearly, path-based flows may be readily translated to corresponding splitting rate values whereas the opposite is not always possible. In fact, awkward values of splitting rates may lead to cyclic vehicle routes; however, proper dynamic traffic assignment algorithms for user optimum or system optimum are not expected to give rise to such unrealistic phenomena that would increase the resulting travel cost.

1.2.3   Model summary

Combining the equations developed in the previous sections, a non-linear discrete-time macroscopic model of the following state-space form

$$\mathbf{x}(k+1) = \mathbf{f}\big[\mathbf{x}(k), \mathbf{u}(k), \mathbf{d}(k), \mathbf{p}\big], \quad \mathbf{x}(0) = \mathbf{x}_0 \qquad (1.20)$$

can be obtained for the entire motorway network, where $\mathbf{x}$ is the state vector, $\mathbf{u}$ is the control vector, $\mathbf{d}$ is the disturbance vector and $\mathbf{p}$ is the parameter vector. This model can be used to simulate the motorway network traffic as shown in Figure 1.3 and to test various control strategies.

In the non-destination-oriented mode, the state vector consists of the densities $\rho_{m,i}$ and the mean speeds $v_{m,i}$ of every segment i of every link m and the queues w_o of every origin and store-and-forward link o. The control vector consists of the ramp metering rates r_o of every origin and store-and-forward link o that is metered. The disturbance vector consists of the demand d_o at every origin o and the turning rates β_n^m at every bifurcation node n.

In the destination-oriented mode, the state vector comprises the partial densities $\rho_{m,i,j}$ of each segment i and reachable destination j from link m, the mean speeds $v_{m,i}$ of every segment i of every link m and the partial queues $w_{o,j}$ of every origin and store-and-forward link o. The control vector consists of the ramp metering rates r_o of every origin and store-and-forward link o that is metered and the splitting rates $\beta_{G,n,j}^m(k)$ at every bifurcation node n where route guidance with respect to destination j takes place. The disturbance vector consists of the demand d_o at every origin o, the

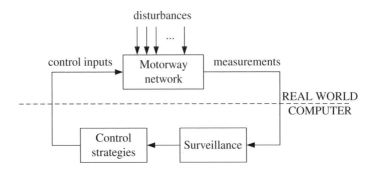

Fig. 1.3 The control loop.

composition rates $\theta_{o,j}$ (OD matrix), the splitting rates $\beta_{n,j}^m$ at every bifurcation node n where no route guidance is applied, the nominal splitting rates $\beta_{N,n,j}^m$ for every bifurcation node n where route guidance with respect to destination j takes place, as well as the drivers' compliance rates.

Finally in both modes, the parameter vector consists of the free speed $v_{f,m}$, the critical density $\rho_{cr,m}$ and α_m for every link m, and the parameters τ, ν, κ, δ and ϕ that are common for all the network links.

This state-space formulation is particularly valuable since it permits the use of well-known methods for the estimation and prediction of the state and parameter vectors as well as for the design of feedback and dynamic optimal control strategies for the motorway network traffic.

1.2.4   Model validation

The model validation procedure aims at enabling the macroscopic model of the motorway network to represent traffic conditions with sufficient accuracy. The estimation of the unknown parameters included in Eq. (1.20) is not a trivial task, since the system equations are highly non-linear in both the parameters and the state variables. The most common approach is the minimization of the discrepancy between the model calculations and the real process in the sense of a quadratic output error functional (Cremer and Papageorgiou, 1981; Papageorgiou et al., 1990; Kotsialos et al., 2002a). For a measurable output vector $\mathbf{y}$ of the nonlinear system Eq. (1.20) given by

$$\mathbf{y}(k) = \mathbf{g}\big[\mathbf{x}(k), \mathbf{p}\big] \tag{1.21}$$

the parameter estimation problem can be formulated as a least-squares output error problem. Given the disturbance and the control vectors, the measured process output $\mathbf{y}^m(k)$ for $k = 0, 1, \ldots, K$, and the initial state $\mathbf{x}_0$ find the set of parameters $\mathbf{p}$ minimizing the cost functional

$$J(\mathbf{p}) = \sum_{k=1}^{K} \big\| \mathbf{y}(k) - \mathbf{y}^m(k) \big\|_Q^2 \tag{1.22}$$

subject to Eqs. (1.20) and (1.21), where $\mathbf{Q}$ is a positive definite, diagonal matrix. The model parameters are selected from a closed admissible region of the parameter space that can be defined on the basis of physical considerations. The determination of the optimal parameter set must be performed by means of a non-linear programming routine. Papageorgiou et al. (1990) demonstrated that the model is most sensitive with respect to the values of the parameters used in the fundamental diagram equation. As these parameters may change in real-time due to varying external conditions (e.g., weather and light conditions, percentage of trucks, variable speed limits, etc.) they could be estimated on-line using the estimation technique described in Section 1.4.

This model validation procedure has been applied by Kotsialos et al. (2002a) to the Amsterdam motorway network shown in Figure 1.4. Each motorway was modeled in both directions as shown in Figure 1.5. The total length of the network was 143 km and its main part was the A10 ring-road that engulfs Amsterdam. Figures 1.6 and 1.7 depict the flow and the speed trajectories, respectively, for link L11 of A10

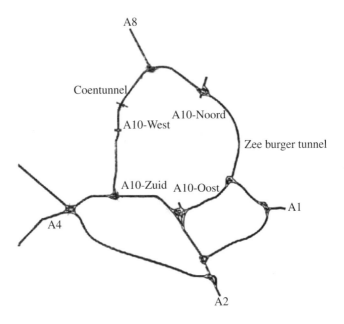

Fig. 1.4 The Amsterdam motorway network.

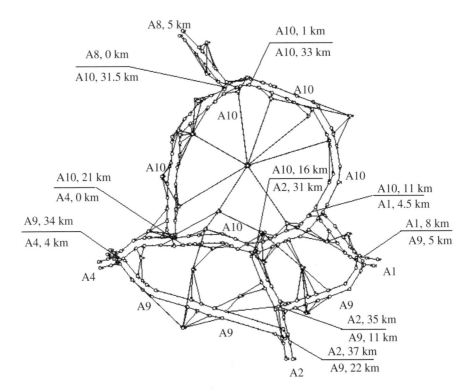

Fig. 1.5 A representation of the Amsterdam network.

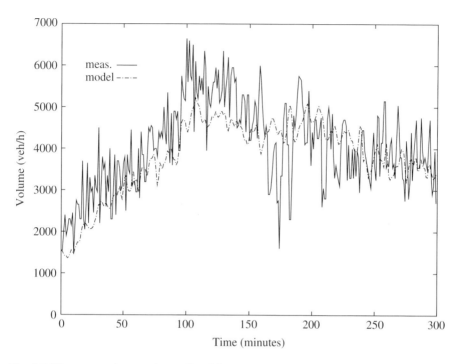

Fig. 1.6 The measured versus the predicted flow.

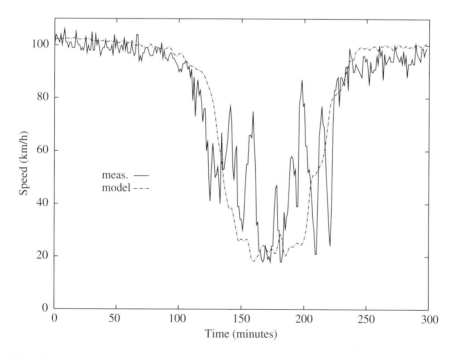

Fig. 1.7 The measured versus the predicted speed.

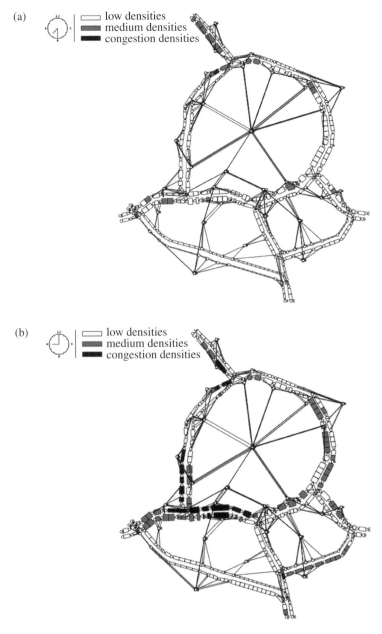

Fig. 1.8 Simulated traffic conditions in the network at (a) 7:30 A.M. and (b) 9:00 A.M.

in the clockwise direction, and compare them to real measurements taken from the same location. Figure 1.8 presents snapshots of the traffic conditions for the whole network at certain time instants. Free, dense and congested conditions are present, and each segment is appropriately colored to indicate the conditions while the segment widths of the links are proportional to the traffic flow passing through them. The model predicted the traffic conditions with sufficient accuracy, which demonstrates that it was a suitable tool for evaluating the impact of various traffic control measures on the traffic flow process.

1.3   DYNAMIC TRAFFIC ASSIGNMENT (DTA) AND ROUTE GUIDANCE (RG)

Motorway networks may include a large number of origins and destinations with multiple paths connecting each origin-destination pair. During rush hours, the travel time on numerous routes changes substantially due to traffic congestion, and alternative routes may become competitive. Drivers who are familiar with the traffic conditions in a network (e.g., commuters) optimize their individual routes based on their past experience, thus leading to user-equilibrium conditions, first proposed by Wardrop (1952). The task of specifying dynamic equilibrium conditions in a traffic network via appropriate routing of sub-flows is known as Dynamic Traffic Assignment (DTA) and is supposed to model the real routing behavior of drivers in an aggregated way. On the other hand, daily varying demands, changing environmental conditions, exceptional events and incidents can alter the traffic conditions in an unpredictable way. This may lead to an underutilization of the overall network's capacity, whereby some links are heavily congested while capacity reserves are available on alternative routes. Route guidance (RG) and driver information systems can be employed to improve the network efficiency via direct or indirect recommendation of alternative routes. Thus, DTA and RG essentially address the same problem (i.e., routing), albeit with different application conditions (off-line modeling versus real-time control).

Routing strategies may aim at either system-optimal or user-optimal traffic conditions. The goal of the former case is the minimization of a global objective criterion (e.g., the total travel time spent) even for the price of partially following routes that are more costly than the regular ones. On the other hand, user-optimal conditions imply equal cost on all utilized alternative routes connecting any two nodes in the network. The algorithms used for DTA or RG can be classified as feedback or iterative. DTA and RG using either feedback or iterative strategies are readily possible with METANET tools (Wang et al., 2001).

1.3.1   Feedback strategies

Particularly for dense networks, with relatively short links, many bifurcations and a large number of alternative routes connecting any two nodes, as well as feedback strategies may be highly efficient in establishing approximate user-optimal conditions on the basis of current traffic measurements. This is particularly appealing in the case of real-time RG. The advantages of feedback strategies include simplicity and low computational effort. Decentralized feedback regulators of P (proportional) or PI (proportional-integral) types have been proposed by Messmer and

Papageorgiou (1994) based on instantaneous (reactive) travel times. Multivariable regulators have also been suggested and can be designed using the linearization of the state Eq. (1.20) around the desired steady-state and Linear-Quadratic (LQ) optimization (Papageorgiou, 1990).

In view of the strongly non-linear character of traffic flow, the closed-loop stability of feedback strategies may be hard to prove formally. Nonetheless, it is usually easy to find controller parameters that lead to satisfactory (albeit approximate) user-optimum results.

1.3.2   Iterative strategies

Iterative strategies may aim at establishing either system-optimal or user-optimal conditions. For the system-optimal case, the procedure is outlined in Section 1.5.2. The typical structure of an iterative procedure towards user-optimal conditions is the following:

(a) Set the initial splitting rates;

(b) Run a simulation model over a time horizon;

(c) Evaluate the experienced (predictive) travel times on alternative routes; if all travel time differences are sufficiently small, stop with the final solution;

(d) Modify the splitting rates appropriately to reduce travel time differences; go to b.

The destination-oriented macroscopic dynamic model developed in Section 1.2.2 can be used in step b. The modification of the splitting rates in step d can be done using decentralized formulas such as the Frank-Wolfe (Wisten and Smith, 1997) and the PI formulas (Wang et al., 2001). The real-time implementation of iterative algorithms for RG purposes employs a rolling horizon procedure in order to reduce the sensitivity with respect to predicted demands and modeling inaccuracies. However, a field implementation of an iterative RG procedure may be a difficult task because of the high computational effort required and the uncertainty created by the need to predict OD, compliance, incidents etc.

Figure 1.9 shows an example network studied by Wang et al. (2001) consisting of 29 nodes and 51 links. For this network, only the (n, j) couples composed of the elements from the sets {N2, N3, N8, N14, N20, and N24} and {D1 and D2} were considered. As there was only one route from N20 to D1, a total of 11 (n, j) couples and hence 11 splitting rates were taken into account. This network includes multiple route choices. Figure 1.10 displays a representative example of routing results. Both diagrams display the splitting rate for the couple (N3,D2) and two relative travel time differences (reactive and predictive). The goal was to keep the travel time differences close to zero by means of a suitable splitting of the related subflows. The simple feedback PI-regulator (Fig. 1.10a) managed to maintain travel time differences fairly low, while an iterative algorithm led to virtually full satisfaction of user-optimum conditions (after 100 iterations), albeit based on perfect knowledge of future demands and driver compliance (Fig. 1.10b).

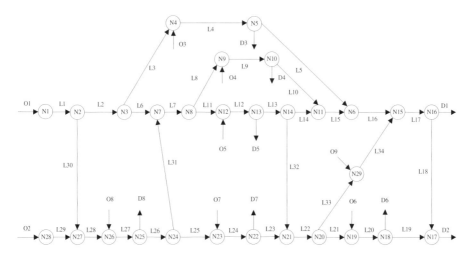

Fig. 1.9 The network topology for the DTA example.

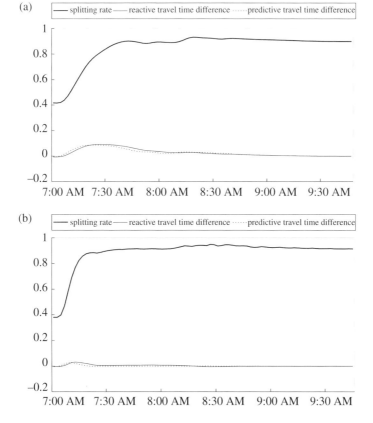

Fig. 1.10 The routing results for node N3 to destination D2 with (a) the feedback strategy; (b) the iterative strategy.

1.4   MOTORWAY NETWORK TRAFFIC SURVEILLANCE

Large-scale motorway networks are usually equipped with several measurement
devices of various kinds (e.g., inductive loops, video sensors, radar detectors, etc.)
delivering real-time information with regard to the current traffic conditions in corresponding locations. However, if the density of available traffic detectors is not very
high (e.g., lower than one sensor per 0.5 or 1 km), the delivered real-time information
may not be complete due to significant space inhomogeneities of the traffic flow.

   Recently, a unified macroscopic model-based approach to real-time motorway
network traffic surveillance along with the REal-time motorway Network trAffIc
State SurveillANCE (RENAISSANCE) tool that has been developed to implement
this approach for field applications was presented by Wang et al. (2004; 2006).
RENAISSANCE is designed on the basis of stochastic macroscopic motorway network traffic flow modeling, extended Kalman filtering and a number of traffic surveillance algorithms. When fed with a limited amount of real-time traffic measurements,
RENAISSANCE enables a number of motorway network traffic surveillance tasks,
including traffic state estimation and short-term traffic state prediction, travel time
estimation and prediction, queue tail/head/length estimation and prediction, model
parameter estimation and incident alarm. RENAISSANCE is a generic tool that is
applicable in real time to motorway networks of arbitrary size, topology and characteristics, based on any suitable traffic detector configurations. Furthermore, it can be
easily integrated in traffic control centers in order to enhance and extend the available
real-time information for various uses.

   A stochastic version of the non-destination-oriented macroscopic dynamic
model developed in Section 1.2.1 represents the staring point of RENAISSANCE,
whereby noise is added on Eqs. (1.2) and (1.3). Moreover, the control inputs **u** may be
considered as additional external inputs within **d**. The state-space form of Eq. (1.20)
is then replaced by

$$\mathbf{x}(k+1) = \mathbf{f}\left[\mathbf{x}(k), \mathbf{d}(k), \mathbf{p}(k), \xi_1(k)\right], \quad \mathbf{x}(0) = \mathbf{x}_0 \qquad (1.23)$$

where **p** is a vector that includes the parameters to be estimated on-line and ξ_1 includes
all the modeling noise. Considering that the disturbance vector $\mathbf{d}(k)$ may not be fully
measurable in real time and that $\mathbf{p}(k)$ is assumed to be unknown, two random-walk
equations are introduced for these parameters

$$\mathbf{d}(k+1) = \mathbf{d}(k) + \xi_2(k) \qquad (1.24)$$

$$\mathbf{p}(k+1) = \mathbf{p}(k) + \xi_3(k). \qquad (1.25)$$

   The combination of Eq. (1.23) and the two random-walk Eqs. (1.24) and (1.25)
leads to the following augmented state-space model

$$\mathbf{z}(k+1) = \mathbf{h}\left[\mathbf{z}(k), \xi(k)\right], \quad \mathbf{z}(0) = \mathbf{z}_0 \qquad (1.26)$$

where $\mathbf{z} = \left[\mathbf{x}^T\ \mathbf{d}^T\ \mathbf{p}^T\right]^T$ and $\xi = \left[\xi_1^T\ \xi_2^T\ \xi_3^T\right]^T$. If noise is added to the measurements,
the output of the system can be expressed as

$$\mathbf{y}(k) = \mathbf{g}\big[\mathbf{z}(k), \eta(k)\big] \tag{1.27}$$

where **y** consists of all the real-time measurements of flow and mean speed, and η includes modeling and measurement noise.

As the model Eqs. (1.26)-(1.27) is highly non-linear, the extended Kalman filter (Jazwinsky, 1970) is employed to design the traffic state estimator according to

$$\hat{\mathbf{z}}(k+1/k) = \underbrace{\mathbf{h}\big[\hat{\mathbf{z}}(k/k-1), \mathbf{0}\big]}_{\text{model}} + \underbrace{\mathbf{K}(k)\big[\mathbf{y}(k) - \mathbf{g}\big[\hat{\mathbf{z}}(k/k-1), \mathbf{0}\big]\big]}_{\text{correction}} \tag{1.28}$$

where $\hat{\mathbf{z}}(k+1/k)$ denotes the traffic state estimation for time instant $k + 1$ based on the traffic measurements available up to time instant k; $\mathbf{K}(k)$ is the gain matrix calculated on-line based on the linear Taylor expansion of **h** and **g** at $\hat{\mathbf{z}}(k/k-1)$. As these calculations are recursive, $\mathbf{K}(k)$ is actually calculated based (implicitly) on traffic measurements at all previous time instants $k - 1, k - 2,...$. Equation (1.28) delivers estimations for the original traffic states **x**, the disturbances **d** and the parameters **p**.

A short-term prediction of the traffic state can be delivered from Eq. (1.23) with the current state estimate used as an initial condition if the impact of unpredictable noise is neglected; short-term predictions are used for the disturbances; and constant parameter values are set equal to the current estimates. The estimation and prediction of travel time are performed on the basis of state estimation and prediction. The reader is referred to Wang and Papageorgiou (2005) for further information on Eq. (1.28) as well as for details on queue tail/head/length estimation/prediction and the incident alarm task.

The following test was conducted on the A92 Motorway close to Munich, Germany in 2004 (Wang et al., 2007). The freeway stretch involved is displayed in Figure 1.11. Five video detector stations (C2, C5, C6, C8, C10), two loop detector stations (L1b and L2), and one radar detector station (R5) were installed along the main stretch, while one loop detector station (L1a) was installed on a motorway merging into the main stretch. Furthermore, two radar detector stations (R3 and R4) were installed at the off-ramp and on-ramp, respectively. Only a subset of these measurements was used to feed RENAISSANCE while the rest of them were utilized solely for evaluating the accuracy of the RENAISSANCE estimates. Figures 1.12a and 1.12b display 24-hour flow and speed measurements collected at C8, C6, and C2 on February 11, 2004. It can be observed in the figures that the speed measurements decreased considerably from midnight until early morning, while the corresponding flow measurements were found to be steady at very low values. Surprisingly, the

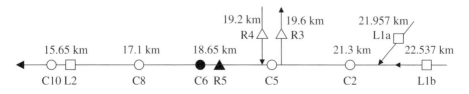

Fig. 1.11 The test stretch on the A92 close to Munich: the stretch layout and detector configuration.

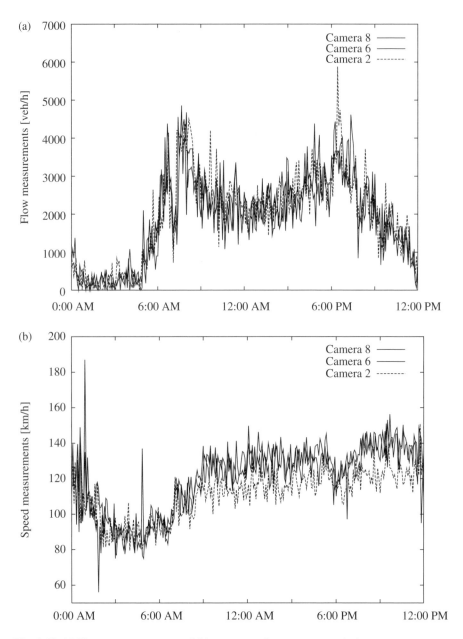

Fig. 1.12 (a) Flow measurements; and (b) mean speed measurements during a snowstorm.

speed measurements during this time period were even lower than those during the afternoon peak period. With the help of the local transportation authority, it was discovered that a snowstorm was present in the area from midnight until the morning of that day. As a result, the free speed of the test stretch decreased substantially, which led to the observed reduction in speed, despite a very low traffic flow.

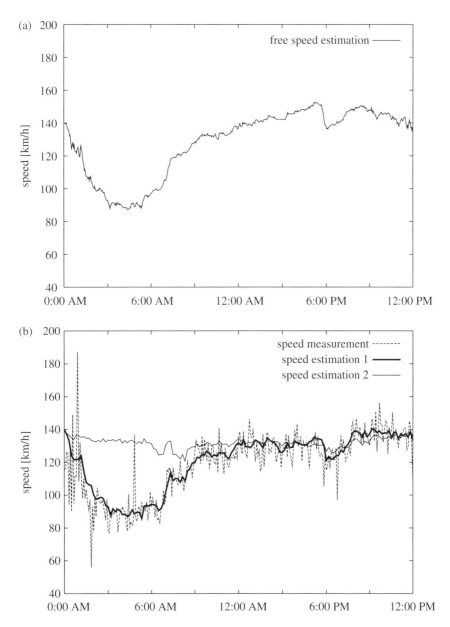

Fig. 1.13 (a) The estimated free speed; and (b) the mean speed estimates at C6 during a snowstorm.

By activating the on-line model parameter estimation, the estimator was able to identify and track the free speed decrease during the snowstorm in real time (Fig. 1.13a) and deliver satisfactory speed estimates at C6 ("estimation 1" in Fig. 1.13b). However, if the free speed was fixed at its usual value of 140 km/h, the estimator failed to track the speed decrease ("estimation 2" in Fig. 1.13b).

1.5   OPTIMAL CONTROL OF MOTORWAY TRAFFIC

The prevention or reduction of traffic congestion on motorway networks can dramatically improve the efficiency of the infrastructure in terms of throughput. Therefore, motorway traffic control represents an increasingly important area in the field of traffic engineering. Local ramp metering strategies and user optimum route guidance may be helpful to a certain extent, but ultimately motorway networks require a superior coordination level that calculates optimal set values from a strategic point of view with the use of optimal control methods and the corresponding numerical solution algorithms. Kotsialos et al. (2002b) presented AMOC, an open-loop control tool that combines a non-linear formulation with a powerful numerical optimization algorithm (Papageorgiou and Marinaki, 1995) and is able to consider coordinated ramp metering, system optimum route guidance as well as integrated control combining both control measures.

1.5.1   Ramp metering

The coordinated ramp metering control problem can be formulated as a dynamic optimal control problem with constrained control variables. It can be solved numerically given demands at every origin of the network and turning rates at the network's bifurcations over a predefined optimization horizon K. For practical applications, these values can be predicted more or less accurately based on historical data or real-time estimations (Wang et al., 2006). The motorway traffic flow is considered as the process under control via the various ramp meters installed at the network entrances. The flow dynamics can be expressed by the non-destination-oriented macroscopic dynamic model developed in Section 1.2.1. The general discrete-time formulation of the optimal control problem is the following:

$$\min J = \vartheta\big[\mathbf{x}(K)\big] + \sum_{k=0}^{K-1}\varphi\big[\mathbf{x}(k),\mathbf{u}(k),\mathbf{d}(k)\big] \qquad (1.29)$$

subject to Eq. (1.20) and the inequality constraints imposed on the ramp metering rates. Here, ϑ and φ are arbitrary, twice differentiable, non-linear cost functions.

The chosen cost criterion is the Total Time Spent (TTS) of all vehicles in the network (including the waiting time experienced in the ramp queues) which is a natural objective for the considered traffic systems. The maximum ramp queue constraints can be taken into account by introducing penalty terms in the cost criterion, thus penalizing queue lengths larger than $w_{max,o}$, i.e., a predetermined maximum admissible number of vehicles in the queue of origin o. Another penalty term can be added in order to suppress high-frequency oscillations of the optimal control trajectories. More precisely, the cost criterion used is the following:

$$J = T\sum_{k=1}^{K-1}\sum_{m}\sum_{i}\rho_{m,i}(k)L_m\Lambda_m + T\sum_{k=1}^{K-1}\sum_{o}w_o(k)$$

$$+ T\sum_{k=1}^{K-1}\sum_{o}\alpha_f\big[r_o(k)-r_o(k-1)\big]^2 + \sum_{k=1}^{K-1}\sum_{o}\alpha_w\big[\max\{0,w_o(k)-w_{max,o}\}\big]^2 \quad (1.30)$$

where α_f and α_w are weighting factors.

The solution determined by AMOC consists of the optimal ramp metering rate trajectories and the corresponding optimal state trajectory and can be used in a rolling horizon manner by applying the controls directly to the network or by employing the state trajectories as set points for local control. Kotsialos and Papageorgiou (2004) gave a detailed presentation of the results from AMOC's application to the problem of coordinated ramp metering in the counter-clockwise direction of the Amsterdam ring-road (Fig. 1.14) over a typical p.m.-peak period. In the considered direction, the motorway A10 has a length of 32 km and includes 21 on-ramps as well as 20 off-ramps. When no ramp metering was applied, the excessive demand coupled with the uncontrolled entrance of drivers into the mainstream, caused a time-space extended congestion, i.e., very high density values in Figure 1.15 blocking almost half of the

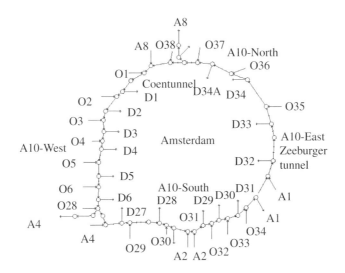

Fig. 1.14 The Amsterdam ring-road.

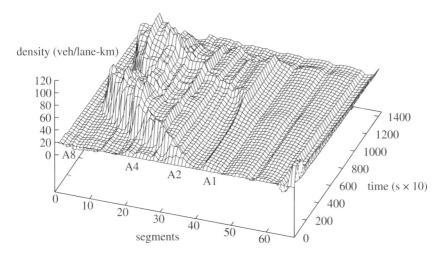

Fig. 1.15 The density profile without control.

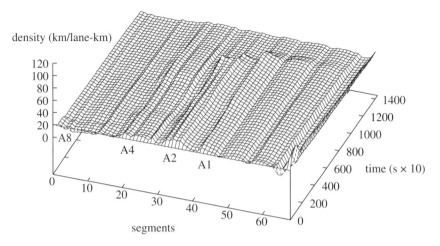

Fig. 1.16 The density profile with optimal ramp metering control.

freeway off-ramps, thus leading to a strongly reduced throughput. By applying the optimal ramp metering, congestion was avoided (Fig. 1.16), the throughput was maximized and the TTS was reduced by 43.5% as compared to the no-control case.

1.5.2   Route guidance

The route guidance problem aiming at establishing system-optimal conditions can be formulated as a dynamic optimal control problem similar to the one presented above. The flow dynamics are now expressed by the destination-oriented macroscopic dynamic model developed in Section 1.1.2 and the control variables are the splitting rates for every bifurcation node where route guidance with respect to a certain destination is applied.

1.5.3   Integrated control

Modern motorway networks may include various types of control measures. The corresponding control strategies are usually designed and implemented independently, thus failing to exploit the synergistic effects that might result from coordination of the respective control actions. The concept of integrated motorway network control may readily result if the dynamic optimal control problem is formulated using the flow dynamics expressed by the destination-oriented macroscopic dynamic model developed in Section 1.1.2. Also required is a control vector including both the ramp metering rates for the controlled on-ramps and the splitting rates for every bifurcation node where route guidance with respect to a certain destination is applied (Kotsialos et al., 2002b).

1.6   CONCLUSIONS

A simple, analytic, discretized, validated traffic flow model in state-space form can provide a unique and convenient basis for efficient handling of a number of significant tasks related to motorway network traffic: simulation, dynamic traffic assignment, traffic state estimation and prediction, real-time parameter estimation, incident alarm, travel time and congestion estimation and prediction, optimal co-ordinated ramp metering, system-optimal or user-optimal route guidance and integrated traffic control.

1.7 REFERENCES

Cremer, M., and Papageorgiou, M. (1981) "Parameter identification for a traffic flow model", *Automatica*, 17:837–843.

Hoogendoorn, S. P., and Bovy, P. H. L. (2001) "State-of-the-art of vehicular traffic flow modelling", *Proceedings of the Institution of Mechanical Engineers*, 215:283–303.

Jazwinsky, A. H. (1970) *Stochastic Processes and Filtering Theory*, Academic Press, New York.

Kotsialos, A., and Papageorgiou, M. (2004) "Efficiency and equity properties of freeway network-wide ramp metering with AMOC", *Transportation Research Part C*, 12:401–420.

Kotsialos, A., Papageorgiou, M., Diakaki, C., Pavlis, Y., and Middelham, F. (2002a) "Traffic flow modeling of large-scale motorway networks using the macroscopic modeling tool METANET", *IEEE Transanctions on Intelligent Transportation Systems*, 3:282–292.

Kotsialos, A., Papageorgiou, M., Mangeas, M., and Haj-Salem, H. (2002b) "Coordinated and integrated control of motorway networks via nonlinear optimal control", *Transportation Research Part C*, 10:65–84.

Messmer, A., and Papageorgiou, M. (1990) "METANET: a macroscopic simulation program for motorway networks", *Traffic Engineering and Control*, 31:466–470.

Messmer, A., and Papageorgiou, M. (1994) "Automatic control methods applied to freeway network traffic", *Automatica*, 30:691–702.

Papageorgiou, M. (1990) "Dynamic modeling, assignment, and route guidance in traffic networks", *Transportation Research Part B*, 24:471–495.

Papageorgiou, M., Blosseville, J. M., and Haj-Salem, H. (1990) "Modelling and real-time control of traffic flow on the southern part of Boulevard Périphérique in Paris – Part I: modeling", *Transportation Research Part A*, 24:345–359.

Papageorgiou, M., and Marinaki, M. (1995) "A feasible direction algorithm for the numerical solution of optimal control problems". Internal Report 1995–4, Dynamic Systems and Simulation Laboratory, Technical University of Crete, Chania, Greece.

Payne, H. J. (1971) "Models of freeway traffic and control", *Simulation Council Proceedings*, 1:51–61.

Wang, Y., Messmer, A., and Papageorgiou, M. (2001) "Freeway network simulation and dynamic traffic assignment with METANET tools", *Transportation Research Record*, 1776:178–188.

Wang, Y., and Papageorgiou, M. (2005) "Real-time freeway traffic state estimation based on extended Kalman filter: a general approach", *Transportation Research Part B*, 39:141–167.

Wang, Y., Papageorgiou, M., and Messmer, A. (2004) "RENAISSANCE – a real-time motorway network traffic state surveillance tool, user manual". Dynamic Systems and Simulation Laboratory, Technical University of Crete, Chania, Greece.

Wang, Y., Papageorgiou, M., and Messmer, A. (2006) "RENAISSANCE – a unified macroscopic model-based approach to real-time freeway network traffic surveillance", *Transportation Research Part C*, 14:190–212.

Wang, Y., Papageorgiou, M., and Messmer, A. (2007) "Investigation of the adaptive features of a real-time freeway traffic state estimator", *Nonlinear Dynamics*, 49:511–524.

Wardrop, J. G. (1952) "Some theoretical aspects of road traffic research", *Proceedings of the Institute of Civil Engineers*, 1:325–362.

Wisten, M. B., and Smith, M. J. (1997) "Distributed computation of dynamic traffic equilibria", *Transportation Research Part C*, 5:77–93.

CHAPTER 2

HYBRID TRAFFIC SIMULATION MODELS: VEHICLE LOADING AT MESO-MICRO BOUNDARIES

Wilco Burghout, Haris N. Koutsopoulos

Traffic simulation models, especially microscopic ones, are becoming increasingly popular and are being used to address a wide range of problems, from planning to operations. However, for applications with large-scale networks, microscopic models are impractical because of input data and calibration requirements. Hybrid models that combine simulation models at different levels of detail have the potential to address these practical issues. This chapter presents a framework for implementing meso-micro hybrid models which facilitates a consistent representation of traffic dynamics. Furthermore, the chapter carries out a detailed examination of an important element impacting the consistent representation of traffic dynamics, i.e., the loading of vehicles from the meso- to the micro-model. A new loading method is presented demonstrating a superior performance as compared to existing approaches. The method is useful not only in the context of hybrid models, but also for microscopic models on their own. A case study illustrates the importance of the method in improving the fidelity of both hybrid and pure microscopic models.

2.1 INTRODUCTION

While microscopic traffic simulation is becoming ever more popular, especially in the evaluation of advanced traffic management systems and intelligent transportation systems (ITS), the amount of effort needed for model calibration and for the preparation of input data often inhibits its use on large networks. Recently, hybrid mesoscopic-microscopic models have appeared (Burghout, 2004; Burghout et al., 2005; Shi and Ziliaskopoulos, 2006; Yang and Morgan, 2006) allowing a detailed microscopic simulation of specific areas of interest, while simulating the remaining areas in lesser detail on a mesoscopic level. Since mesoscopic simulation has a more aggregated representation of the roadway and the vehicle interactions, it requires much less effort in calibration and preparation of input data (especially coding of the road network). In addition, a number of hybrid macroscopic-microscopic models have recently appeared (Bourrel and Lesort, 2003; Espie et al., 2006; Magne et al., 2000; Mammar et al., 2006; Poschinger et al., 2000).

The development and implementation of hybrid models that combine traffic simulation models at different levels of detail require the resolution of a number of issues, some of them related to the interaction of the two models at their boundaries. Among these issues, the consistency in traffic dynamics at the meso- and micro-network boundaries is particularly important.

The objective of this chapter is twofold: (a) to present a general framework for the implementation of hybrid simulation models satisfying the various integration requirements; and (b) to detailed discuss an important aspect that has serious implications for the validity of simulation models in general and hybrid models in particular. This aspect is the mechanism used to load vehicles arriving from the mesoscopic area into the microscopic area, which affects the consistency of traffic dynamics at the meso-micro boundaries.

The remainder of the chapter is organized as follows. Section 2.2 discusses important requirements for hybrid models of high fidelity and presents a general framework for their implementation. Section 2.3 looks into modeling traffic dynamics at the micro-meso boundary in more detail. Section 2.4 addresses issues related to existing vehicle loading mechanisms and proposes a loading mechanism that alleviates the negative impact of prior methods. The latter problem is important not only in the context of hybrid simulation models but also microscopic simulation models on their own. Section 2.5 illustrates the impact of the loading mechanism through a case study with a small network and Section 2.6 concludes the chapter.

2.2   HYBRID MODELING FRAMEWORK

The main requirements that the integration of micro/meso models needs to satisfy in order to develop reliable hybrid models include:

- *Consistency in network representation.* One of the most basic conditions is the general consistency of the two models in their representation of the road network, especially at the boundaries between the two models. A consistent network representation has important implications for capacity determination, and hence impacts traffic dynamics at the boundaries between the models. Furthermore, it impacts the links included in each model and hence, the consistency with respect to route choice as discussed below.

- *Consistency in route choice representation.* One of the most important conditions is the consistency of the two models in their representation of paths, and route choice alternatives. The route choice needs to be consistent across the models to ensure that vehicles will make the same decision given the same route choice situation (pre-trip or en-route), regardless of if they are in the micro- or the meso-model. This also means that the representation of the alternative paths needs to be consistent throughout the hybrid model, as do the travel times (link costs).

- **Consistency of traffic dynamics at meso-micro boundaries.** Besides the consistency of network representation, the consistency of traffic dynamics at the boundaries between the meso- and micro-submodels needs to be ensured. In other words, the traffic dynamics upstream and downstream of the boundaries must be consistent. For instance, when a queue forms downstream of the boundary point, and grows until it reaches the boundary, it should continue in the other submodel, upstream the boundary, similarly to how it would have done if the boundary had not been there.

- **Consistency in traffic performance for meso and micro submodels.** The two submodels need to be consistent with each other with regard to the results they produce. Ideally, for the facilities that can be simulated sufficiently well by both models, the results in terms of common outputs such as travel times, flows, speeds, densities, etc, should be similar. This implies the need for a consistent calibration of the two models.

- **Transparent communication and data exchanges.** The submodels exchange large amounts of data conveying vehicle characteristics and downstream traffic conditions. This requires an efficient synchronization and communication paradigm and a design that minimizes the amount and frequency of data exchange. Otherwise, the communication overhead may become very large. On the other hand, aggregation and disaggregation of information at the boundaries may introduce complications and should therefore be avoided.

Figure 2.1 shows a general hybrid simulation framework (Burghout et al., 2005) that facilitates the resolution of the issues discussed above. The travel behavior component in the common module uses the database that contains the complete network representation, link travel times and known paths for the entire network. Both

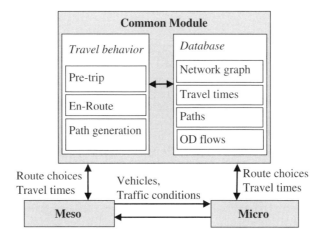

Fig. 2.1 A hybrid simulation framework.

the micro and meso models supply descriptions of their subnetworks, from which the network representation is constructed. Each time a vehicle makes a route choice (whether in micro or meso), the common module is invoked. For the common route choice module to operate properly, the meso and micro models need to update the travel time database regularly with the link travel times in their subnetworks.

The common module also includes the OD matrix for the entire network. To simplify the information exchange and facilitate a transparent data input interface, it is assumed that all origin and destination nodes in the network belong to the meso-sub-network. This assumption is not restrictive, since an origin or destination node in the micro area can always be designated as a boundary node in meso, connected directly to the micro-subnetwork. In addition, the meso-model contains the entire network and route representation, by representing the paths inside the micro, from entry nodes to exit nodes, as *virtual links*. These virtual links are then used as part of the meso-network description, in order for the meso-model choice to provide consistent route choices across the submodels. This issue is discussed in more detail in Burghout, 2004; Burghout et al. (2005).

2.3   MODELING TRAFFIC DYNAMICS AT MESO-MICRO BOUNDARIES

Consistency with respect to traffic dynamics at the boundaries of the two models is critical. The limited literature on hybrid simulation models focuses on aggregation/disaggregation issues between the macro and micro representations of traffic (Bourrel and Lesort, 2003; Espie et al., 2006; Magne et al., 2000; Mammar et al., 2006; Poschinger et al., 2000). In the case of meso/micro models, these issues are avoided since the integration involves two models that both have a vehicle-based representation of traffic flows. However, other issues concerning the interfaces between the models remain. In principle, the main sources of potential inconsistencies at the interface between the meso- and micro-models involve:

- location of boundaries;

- queue formation and representation;

- vehicle attributes (i.e., the determination of speeds and accelerations of vehicles crossing the micro/meso boundaries).

Location of boundaries. Due to the fact that the meso- and micro-submodels may provide differing representations of links and intersections the positioning of the boundaries between them should be carefully considered. Placing them at nodes in the overall network would imply that some of the legs of an intersection would be at a meso- and other at a micro-level. This creates the problem of having to deal with the intersection behavior (signal-controlled or otherwise) at various levels of detail. For instance, gap acceptance models in the micro-module require vehicles' positions, speeds and gaps in the opposing flows. However, this information is not typically available at the mesoscopic level. It is therefore recommended that the boundaries be located in the middle of the links, with exactly one entry and one exit segment.

Queue formation and representation. The placement of boundaries is also dependent on the homogeneity of traffic conditions. Typically, micro-models, due to their level of detail, represent lane-specific queues. Meso-models, on the other hand, usually do not have lanes. If this is the case, a queue on one lane in micro will block the boundary node completely, even if the other lanes remain open, unless the exchange of vehicles at the boundary takes place properly. The virtual links in the meso-subnetwork play an important role in solving the problem of inconsistent queue representation. Virtual links representing paths that can only use the obstructed lanes will be blocked, while virtual links representing paths that use open lanes will have an available capacity and allow the proper advancement of the corresponding vehicles.

Crossing vehicle attributes. The attributes of the vehicles as they cross the boundaries need to be properly determined in both directions (micro → meso) and (meso → micro). Attributes such as speeds, accelerations, headways, etc. should be consistent with the prevailing conditions in the new segment that the vehicle is moving to, otherwise unnecessary shockwaves may propagate upstream. On the boundary from the meso to the micro (meso → micro) submodel, information is exchanged in both directions: from meso to micro, information concerning vehicles (with a certain speed and at certain time intervals) needs to be communicated; from micro to meso, information regarding the blocking of boundaries and downstream density needs to be conveyed. If the entry to the micro-link (downstream of the boundary point) is blocked, the meso needs to prevent vehicles from exiting. The micro-model informs meso when the obstruction is removed, so that vehicles can start flowing (over that specific boundary) again. The micro also sends the density in the vicinity of the boundary to meso, where it is used to calculate the speed of the shockwave that propagates upstream.

One very important aspect that relates to the questions raised above is the loading of vehicles from the meso- to the micro-model. Vehicle loading in hybrid and microscopic simulation models comprises the generation of *initial values* for state *variables* such as headway and speed. This is different from the correct setting of *parameters* such as the desired speed, which consists of static inputs, whereas state variables vary continuously as they describe the state of the vehicles.

2.4   VEHICLE LOADING

When vehicles enter the microscopic network, the initial values for speed and headway need to be in accordance with each other as well as with the traffic situation downstream (and upstream). Otherwise, an immediate deceleration (or acceleration) will occur, affecting the traffic conditions upstream. A correct vehicle loading is crucial in a hybrid model since the artificially created disturbances propagate upstream and cause shockwaves in the mesoscopic model. It is equally important for standalone microscopic simulation models, since it affects the capacity at the entrance of microscopic networks. This is all the more urgent since, in most applications of microscopic models, the vehicles are loaded at links, rather than at centroids.

The issue of vehicle loading in microscopic models has been largely neglected until now, and publications and documentation of microscopic traffic models provide little or no information on how this issue is addressed. The fact that the loaded vehicles are at the entrance of the network makes it difficult to observe errors in the loading mechanism since their immediate deceleration (or acceleration) does not cause an observable disturbance behind them. However, as will be shown in this chapter, these vehicles do cause a drop in capacity at the entrances to the network, and may therefore affect the simulation results by prohibiting the loading of traffic at high volumes, or even cause distortion in the calibration process if other parameters are adjusted to compensate for the erroneous loading procedure. In short, incorrect vehicle speeds at micro-meso boundaries or micro-origins lead to:

- excessive decelerations and accelerations;

- volatility and delayed vehicle entry;

- reduced capacity at entry points;

- unnecessary turbulence and shockwave propagation upstream the meso-component of hybrid models.

2.4.1   Vehicle loading in existing microscopic models

Although details regarding the loading mechanism in existing micro-simulation models are sketchy, the present chapter describes the default loading method used in two state-of-the-art models, MITSIMLab (Yang et al., 2000), and VISSIM (PTV, 2008). *MITSIMLab.* The vehicle loading in MITSIMLab consists of lane selection and speed assignment components. The lane selection depends on lane permissions (e.g., bus lanes), the continued path of the vehicle (in order for it not to have to immediately change lanes to continue its path) and available space on lanes. Of the 'acceptable lanes', the lane with the most space is selected. The speed of the loaded vehicle is set based on the vehicle's desired speed, the average segment speed and the entry queue length (see Fig. 2.2). If the entry queue is 0 (i.e., no vehicles waiting to enter), the vehicle is loaded with its desired speed. When the entry queue grows (due to vehicles not being able to enter at the rate they arrive), the vehicle is loaded at a speed approaching the segment speed. The problem that this method may cause is that, in congested situations, vehicles may be loaded at much higher speeds than the vehicles in front, causing immediate deceleration. This is expected to give rise to turbulence and shockwaves, and result in a reduced capacity for entry links. *VISSIM.* As in MITSIMLab, the lane selection in VISSIM depends on lane permissions and the available space headway, where the acceptable lane with the largest space headway is selected. The speed of the loaded vehicle in VISSIM is a function of the vehicle's desired speed, the speed of the vehicle in front (on the selected lane) as well as the space headway to this vehicle. If the front vehicle is beyond the look-ahead distance of the vehicle to be loaded, the speed of the loaded vehicle will constitute its desired speed. If the distance to the vehicle in front is 0, the vehicle would, in theory, be loaded with the same speed as the vehicle in front (provided that the vehicle in front has a lower speed than

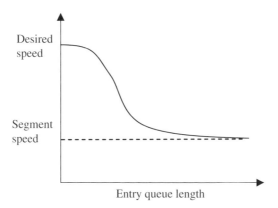

Fig. 2.2 The vehicle loading speed in MITSIMLab.

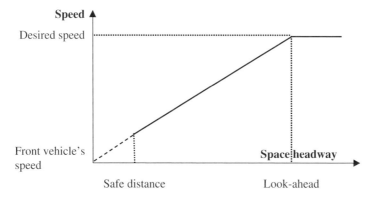

Fig. 2.3 The vehicle loading speed in VISSIM.

the loaded vehicle's desired speed). However, the vehicle to be loaded will not be able to enter the network if the space headway is smaller than a pre-defined safety distance. Thus, in practice, the vehicle will usually be loaded at a speed somewhat higher than that of the front vehicle. While this method will prevent extreme decelerations at the entry, it will still induce slight shockwaves due to the fact that, in saturated conditions, most entering vehicles have a speed that is somewhat higher than the speed of the vehicle in front and therefore always decelerate upon entry (see Fig. 2.3).

2.4.2 Proposed loading method

Here, a new loading method is proposed, aiming at loading vehicles in a way that is more consistent with their behavior when they enter the network. In microscopic simulation models, the car-following model is one of the main models used to move vehicles in the network. Typical car-following models relate the acceleration of a vehicle to the behavior of the surrounding vehicles, especially the speed of, and the distance to, the leading vehicle. The proposed method assigns speed to the vehicles entering the network (either loaded at the origin, or transferred from the mesoscopic part of the network) using a multi-regime approach based on the time headway between the entering vehicle and its leading vehicle (in the same lane):

Regime 1 (bound traffic): $t_1 < t_h \leq t_2$ seconds $V = V_{front}$

Regime 2 (partially bound traffic): $t_2 < t_h \leq t_3$ seconds

$$\alpha = (t_h - t_2)/(t_3 - t_2)$$

$$V = \alpha * V_{desired} + (1 - \alpha) * V_{front}$$

Regime 3 (unbound traffic): $t_h > t_3$ seconds

$$V = V_{desired}$$

where

V	:	assigned initial speed;
V_{front}	:	speed of the vehicle in front (in the same lane);
$V_{desired}$	:	vehicle's desired speed;
t_h	:	time headway to the vehicle in front (in the same lane);
α	:	parameter [0,1] that interpolates between the desired speed and the speed of the leading vehicle. If $V_{front} > V_{desired}$, $V = V_{desired}$;
t_i	:	time headway thresholds, determined/calibrated from field data, $i \bullet \{1, 2, 3\}$.

The time headway from the leading vehicle is defined as the time that has passed since the vehicle in front was at the loading point, *assuming it drove at its current speed*. This means that, in case of queue build-up and propagation towards the entry point, the vehicle in front is likely to have decelerated since entering, and therefore the time headway may be slightly overestimated.

The above loading mechanism is supported by empirical data. On the E4 motorway (Essingeleden) in Stockholm, individual speed and time headway data was collected (by the Swedish National Road Administration) with high fidelity microwave detectors, using one detector for each lane. During the measurement period, the flows varied between 800 and 1700 veh/h/lane.

The measured speed and time headway data was grouped into time headway classes of 1 second each, i.e., one class for vehicles arriving with time headways of 0-1 seconds, another class for 1-2 seconds etc. Figure 2.4 illustrates the correlation coefficient between speeds of consecutive vehicles as a function of their time headway. A clear trend can be observed from the data, with a very high correlation for the vehicles following at close time headways (classes 0-1, 1-2, and 2-3 seconds). As expected, for the 3-4, 4-5 and 5-6 second time headway classes, the correlation coefficients decreased. It should be noted that the sample size of the classes for longer time headways was much smaller than for the classes for short headways.

Figure 2.5 compares the actual speeds of two consecutive vehicles in the two extreme classes, i.e., 1-2 seconds and 7-8. The scatter plots clearly show the high

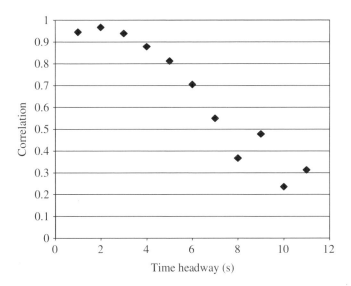

Fig. 2.4 The consecutive vehicle speed correlation as a function of the time headway.

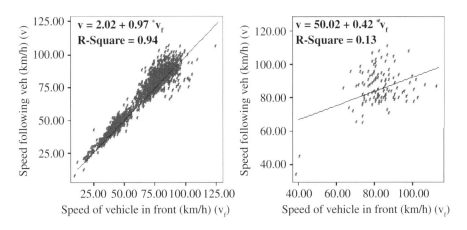

Fig. 2.5 Scatter plots of consecutive vehicle speeds: (a) 1-2 second group; (b) 7-8 second group.

correlation in the 1-2 second group, as compared to the (mostly absent) correlation in the 7-8 second group.

Based on the above data, the following default headway thresholds for the proposed loading method were estimated as follows:

- $t_1 = 0.5$ seconds (minimum headway parameter);

- $t_2 = 2.5$ seconds;

- $t_3 = 7.5$ seconds.

2.5   CASE STUDY

The importance of the loading can be demonstrated by using two state-of-the-art microscopic traffic simulation models, MITSIMLab (Yang et al., 2000) and VISSIM (PTV, 2008). The impact on the performance of a hybrid model can also be shown by the MiMe model (Burghout, 2004) that combines MITSIMLab and the mesoscopic model MEZZO (Burghout, 2004).

A small, linear network was used to demonstrate the impact of the loading mechanism. The network in Figure 2.6 evaluated the performance of each simulation model (MITSIMLab and VISSIM).

The network is loaded with a time dependent demand that increases slowly from $t = 0\text{-}1800$ seconds and creates relatively high flows from $t = 1800\text{-}2400$ seconds, after which it gradually decreases (Fig. 2.7).

The importance of the loading method is best illustrated by the distribution of speeds at sensor locations 0 (at the entry) and 1 (500 m downstream). Figure 2.8 shows the speed at these locations for (a) MITIMSLab and (b) VISSIM when the default loading mechanism was used.

The loading in MITSIMLab clearly indicated a number of problems, especially since the vehicles entered the network at rather high speeds. Although the speeds in VISSIM appeared more consistent between the sensors, problems still existed with respect to the loading of vehicles in VISSIM. Figure 2.9 shows the minimum acceleration (i.e., the maximum deceleration in absolute values) of the vehicles entering the network in VISSIM. Ideally, the minimum accelerations over sensors 0 (at entry) and sensor 1 (after 500 m) should be similar. However, Figure 2.9 illustrates that the

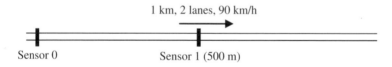

Fig. 2.6 Test network for MITSIMLab and VISSIM alone.

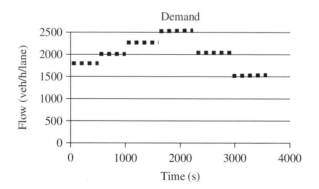

Fig. 2.7 The demand profile.

(a)

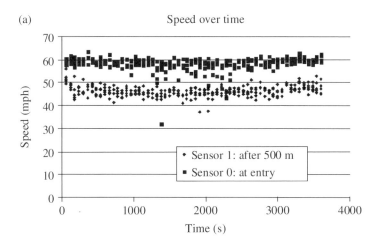

(b)

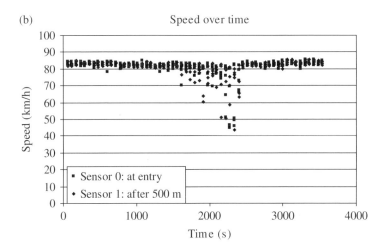

Fig. 2.8 The impact of the default loading mechanism: (a) MITSIMLab; (b) VISSIM (80 km/h = 50 mph).

minimum accelerations over sensor 0 were much greater than those over sensor 1, suggesting that some friction was caused by the vehicle loading. This has implications on the number of vehicles that the simulator is able to load on the network (capacity at entry points). During the peak demand period (1800-2400 seconds), 18 vehicles were unable to enter the network.

The proposed loading mechanism was implemented in MITSIMLab. Figure 2.10(a) shows the speeds of the vehicles over sensors 0 and 1 with the new loading mechanism. Figure 2.10(b) shows the corresponding average accelerations during the first 20 seconds after the vehicles entered the network (at 0.1-second intervals), and compares them to the acceleration of the vehicles with the original loading method.

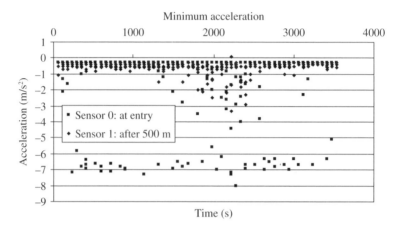

Fig. 2.9 The minimum acceleration of vehicles in VISSIM.

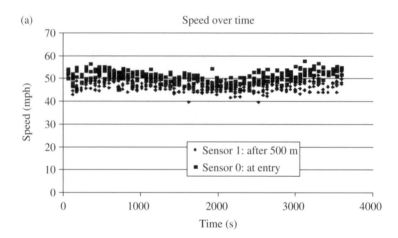

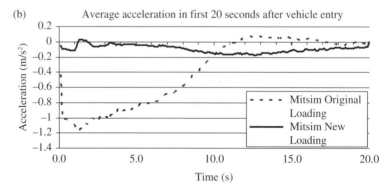

Fig. 2.10 MITSIMLab: the impact of the proposed loading mechanism: (a) speeds with improved loading; (b) average accelerations.

As discussed before, the original loading method in MITSIMLab generated initial speeds (sensor 0) that were too high for the time headway and the traffic conditions immediately downstream (see Fig. 2.8a). This caused large decelerations of the entering vehicles (see Fig. 2.10b), and due to the resulting turbulence, the capacity of the entry was underestimated. In contrast, the proposed loading method demonstrated a much better behavior. The average speeds at the link entry were comparable to those in the interior of the link (Fig. 2.10a), causing little deceleration upon entry (Fig. 2.10b). Consequently, the flows were no longer artificially disrupted by the decelerations of the entering vehicles (Fig. 2.10b).

In addition, the impact of the loading mechanism was assessed using the hybrid simulation model MiMe in the same network. The mesoscopic (MEZZO) and microscopic (MITSIMLab) areas of the network are shown in Figure 2.11. MiMe was first used to simulate the operations of the network with the default MITSIMLab loading mechanism, and subsequently with the proposed method.

The results displayed that the original loading method caused disturbances resulting in high accelerations (Fig. 2.12). This behavior was similar to that observed with MITSIMLab alone, although the impact seemed to be greater in the hybrid model (compare Figs. 2.10b and 2.12).

The flows, under the default loading method, were very unstable (Fig. 2.13a). The decelerations gave rise to heavy disruptions of the flows, due to the shockwaves that propagated backward into the mesoscopic network. The proposed loading method generated more realistic accelerations during the first 20 seconds (Fig. 2.12) and, as a result, a much better flow profile since the flows were no longer artificially disrupted by the decelerations of entering vehicles (Fig. 2.13b).

Fig. 2.11 The test network for the MiMe hybrid model.

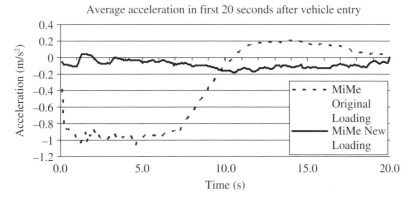

Fig. 2.12 The entry acceleration pattern with the original and proposed loading method in MiMe.

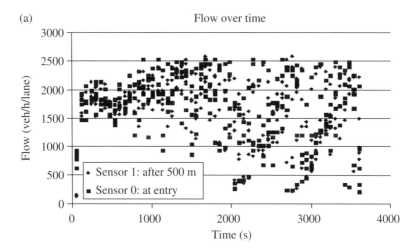

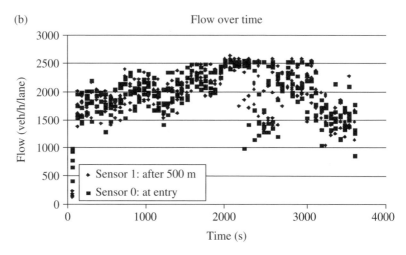

Fig. 2.13 Flows in the MiMe hybrid model. (a) The original loading method. (b) The proposed loading method.

2.6   CONCLUSIONS

The present chapter has discussed issues related to the integration of mesoscopic-microscopic traffic simulation models. Such hybrid models have a number of advantages over standard microscopic models, especially in large scale applications with extensive data input and calibration requirements. Successful hybrid models require, among other things, a consistent representation of traffic dynamics at the boundaries between the meso- and micro-networks. The loading mechanism represents a critical determinant of this consistency as it transfers vehicles form the meso- to the micro-network. The loading mechanism was found to be important for both hybrid as well as stand-alone micro-models.

This chapter has presented an integrated framework for the development of hybrid models. In addition, the role of the vehicle loading mechanism on the validity of hybrid models (and microscopic models on their own) was also examined. A new loading method was proposed which, as compared to existing approaches, gave rise to a much more consistent performance. The results from a case study supported both the importance of the loading mechanism and the value of the proposed method. Inappropriate loading methods can result in excessive decelerations and accelerations, volatility and delayed vehicle entry, reduced capacity at entry points, and unnecessary turbulence and shockwave propagation.

2.7   REFERENCES

Bourrel, E., and Lesort, J.-B. (2003) *Mixing Micro and Macro Representations of Traffic Flow: A Hybrid Model Based on the LWR Theory,* Washington DC, Transportation Research Board.

Burghout, W. (2004) "Hybrid microscopic-mesoscopic traffic simulation", *Department of Infrastructure.* Stockholm, Royal Institute of Technology: 210. http://www.infra.kth.se/ctr/publikationer/ctr2004_04.pdf.

Burghout, W., Koutsopoulos, H., and Andreasson, I. (2005) "Hybrid mesoscopic-microscopic traffic simulation", *Transportation Research Record*, 1934:218–225.

Espie, S., Gattuso, D., and Galante, F. (2006) *A Hybrid Traffic Model Coupling Macro and Behavioural Micro Simulation*, 85th Meeting of Transportation Research Board CD-ROM, Washington DC.

Magne, L., Rabut, S., and Gabard, J.-F. (2000) *Towards An Hybrid Macro-Micro Traffic Simulation Model*, INFORMS, Salt Lake City, USA.

Mammar, S., Lebaque, J.-P., and Haj-Salem, H. (2006) *A Hybrid Model Based on a Second Order Traffic Model*, 85th Meeting of Transportation Research Board CD-ROM, Washington DC.

Poschinger, A., Kates, R., and Meier, J. (2000) *The Flow of Data in Coupled Microscopic and Macroscopic Traffic Simulation Models*, World Congress on Intelligent Transportation Systems, Torino.

PTV (2008) *VISSIM User's Manual 5.1*, PTV AG, Karlsruhe.

Shi, H., and Ziliaskopoulos, A. (2006) *A Hybrid Mesoscopic-Microscopic Traffic Simulation Model: Design, Implementation and Computational Analysis*, 85th Meeting of Transportation Research Board, CD-ROM.

Yang, Q., Koutsopoulos, H. N., and Ben-Akiva, M. (2000) "A simulation laboratory for evaluating dynamic traffic management systems", *Transportation Research Record*, 1710:122–130.

Yang, Q., and Morgan, D. (2006) *A Hybrid Traffic Simulation Model*, 85th Meeting of the Transportation Research Board CD-ROM, Washington DC.

CHAPTER 3

SIMULATION OF VEHICLES IN A DRIVING SIMULATOR USING MICROSCOPIC TRAFFIC SIMULATION

Johan Janson Olstam

This chapter describes a model that generates and simulates surrounding vehicles for a driving simulator. The proposed model generates a traffic stream, corresponding to a given target flow and simulates realistic interactions between vehicles. The model is built on established techniques for a time-driven microscopic simulation of traffic and uses an approach of only simulating the closest neighborhood of the driving simulator vehicle. In the presented model, this closest neighborhood is divided into one inner region and two outer regions. Vehicles in the inner region are simulated according to advanced sub-models for driving behavior, while vehicles in the outer regions are updated according to a less time-consuming model. The presented work includes a new framework for generation and simulation of vehicles within a moving area. It also includes the development of an enhanced model for overtakings and a simple mesoscopic traffic model. The developed model has been tested within the VTI Driving simulator III. A driving simulator experiment has been conducted in order to verify whether the participants observe the behavior of the simulated vehicles as realistic or not. The results were promising but also indicated that enhancements could be made. Moreover, the model was validated on the number of vehicles that caught up with the driving simulator vehicle and vice versa. The agreement was good for active and passive catch-ups on rural roads and for passive catch-ups on freeways, but less good for active catch-ups on freeways.

3.1 INTRODUCTION

Numerous traffic accidents are caused by failures in the interaction between the driver, the vehicle, and the traffic system, which is why knowledge of such interactions is essential. Nowadays, this is especially true since the number of driving-related interactions is increasing. Drivers of today also interact with advanced driver-assistance systems, in-vehicle information systems, various nomad-devices, such as mobile phones, personal digital assistants, etc, and these technical systems influence drivers' behavior and their ability to maneuver a vehicle.

To obtain this knowledge, researchers conduct behavioral studies and experiments, performed either in real traffic systems, on a test track, or in a driving simulator. The real world is of course the most realistic environment, but can be unpredictable with respect to for instance weather, road and traffic conditions. For this reason, it is often hard to design real-world experiments from which it is possible to draw statistically significant conclusions. Some experiments are also too dangerous or impossible to conduct due to ethical reasons. Test tracks offer a safer environment and the possibility of giving test drivers more equal conditions, but they also have drawbacks, notably regarding the variety and complexity of the driving context. Driving simulators, on the other hand, offer a less realistic environment than that of real-world experiments, but in which test conditions can be fully controlled and varied in a safe manner.

A driving simulator is designed to imitate driving a real vehicle. The driver's place can be realized with a real vehicle cabin or only a seat with a steering wheel and pedals, and anything in between. The surroundings are presented for the driver on a screen. It is important that the performance of the simulator vehicle, the visual representation, and the behavior of surrounding objects are realistic in order for the driving simulator to be a faithful representation of real driving. It is for instance clear that the surrounding vehicles must behave in a realistic and trustworthy way. Microscopic simulation of traffic is one possibility for simulating these surrounding vehicles. Micro-models use various sub-models for car-following, lane-changing, speed adaptation, etc. to simulate driver behaviors at a microscopic level. The present chapter concerns the simulation of surrounding vehicles in driving simulators using traditional techniques for microscopic simulation of traffic.

The main reason for choosing driving simulators for conducting driving behavior experiments is often to obtain an increased controllability and reproducibility. In order to ensure a high reproducibility, the behavior of the surrounding road-users is often strictly controlled. This comes at a price of limited realism regarding the behavior of surrounding vehicles and the limitations in the complexity of the scenario situations, due to both the complexity of the scenario programming and the required programming effort. The complexity of programming can be decreased and the realism increased by providing the surrounding road-users with more autonomy. When using autonomous surrounding traffic, drivers will experience different situations at the micro-level depending on how they drive. The simulator drivers' conditions will still be comparable at a higher, more aggregated, level. This leads to simulated situations with similar pros and cons as in real world situations, i.e. a low reproducibility but realistic surroundings. However, this environment is both safer and remains more controllable than the real world. Using autonomous surrounding traffic also makes it much easier to conduct driving simulator experiments with dense traffic conditions.

Some trials in which commercial software packages such as AIMSUN (Barceló and Casas, 2002) and VISSIM (PTV, 2003) were used to simulate surrounding vehicles in driving simulators have been conducted, see for example (Bang and Moen, 2004) and (Jenkins, 2004). However, since commercial programs simulate a specified geographic area, very large areas have to be simulated when running long driving simulator experiments (i.e. 1–2 hours of driving). A more efficient approach would therefore be to only simulate the closest neighborhood of the driving simulator vehicle. Naturally, this area moves with the same speed as the simulator vehicle. Unlike most applications of traffic simulation, the important output for this application is at the micro level, i.e. the actual behavior of the simulated vehicles/drivers. Traffic simulation models often include assumptions and simplifications that do not affect the model validity at the macro level but sometimes influence the validity at the micro level. One typical example is the modeling of lane-changing movements. In most simulation models, vehicles change lanes instantaneously. This is not very realistic from a micro-perspective but does not affect macro measurements appreciably. The present application therefore requires a more detailed modeling of vehicle movements than that used in most software. Another reason for not using available commercial software is their inability to simulate rural roads with oncoming traffic, or the fact that the modeling of such roads is not detailed enough for this kind of application. In the light of these facts, a couple of models specialized for the application of simulation of surrounding vehicles in driving simulators has been developed. Three of the most well-known models are the ARCHISIM model (El Hadouaj and Espié, 2002; Espié, 1995), the NADS model (Ahmad and Papelis, 2001), and the DRIVERSIM model (Wright, 2000). Other interesting work within this area includes e.g. that of (Bayarri et al., 1996; Champion et al., 1999; Kuwahara et al., 2005). Research within this area has to a large extent been directed toward decision-making modeling concepts. Focus has also to a considerable degree been limited to the simulation of freeways. There has been little interest in the modeling of rural roads and in algorithms for the generation of realistic traffic streams.

The objective of this project was to develop, implement, and validate a real-time running traffic simulation model capable of generating and simulating surrounding vehicles in a driving simulator. This includes the integration of the developed model and a driving simulator. The model should simulate both individual vehicle-driver units and the traffic stream that they are a part of in a realistic way. The simulated vehicle-driver units should behave realistically concerning acceleration, lane-changing, and overtaking behavior, as well as with regard to speed choices. The vehicles should also appear in the traffic stream in such a way that headways, vehicle types, speed distributions, etc. correspond to real data. The simulation model has been delimited to only deal with freeways with two lanes in each direction and to rural roads with oncoming traffic. The model does not address ramps on freeways or intersections on rural roads.

The developed simulation model including the general simulation framework, methods for the generation of vehicles, and sub-models for driving behavior is presented in Section 3.2. In Section 3.3 follows a presentation of the two performed validation studies. Finally, Section 3.4 ends with some concluding remarks and suggestions for future research.

The main contribution of this chapter is the general simulation framework, the technique for generation of new vehicles, the enhanced overtaking model, and the mesoscopic simulation model.

3.2   THE SIMULATION MODEL

The simulation model is based on established techniques for a time-driven micro-simulation of road traffic. The model simulates surrounding traffic corresponding to a given target traffic flow and traffic composition. Moreover, it uses the simulator vehicle's speed, position, etc. as input and generates the corresponding information concerning the surrounding vehicles as output.

As in most traffic simulation models, vehicles and drivers are treated as one unit, i.e. the interaction between the driver and the vehicle is not modeled. These vehicle-driver units are described by a set of driver or vehicle characteristics, including the desired speed, the desired following time gap, the power/mass-ratio, etc. Both vehicle and driver characteristics vary between different vehicle types. At the moment, the model includes the following vehicle types: Cars, Buses, Trucks, Trucks with a trailer with three or four axles, and Trucks with a trailer with five or more axles. However, buses and trucks without trailers are assumed to have equal characteristics.

The simulation model follows a traditional time-discrete update approach. The update procedure has been divided into two parts: in the first part, the speed and position are updated for all vehicles, and in the second part the behavior of the simulated vehicles is updated. This refers to acceleration, lane-changing and overtaking decisions, etc. By separating the position and behavior updating, it is possible to avoid that information of already updated vehicles is used to update the behavior of the rest of the vehicles.

3.2.1   The moving window

Instead of simulating a specific geographical area, an approach where only the closest neighborhood of the simulator vehicle is simulated is employed. Similar approaches have been used in the ARCHISIM model (Espié, 1995) and in the NADS model (Bonakdarian et al., 1998). This neighborhood or area moves with the same speed as the simulator vehicle and can be interpreted as a moving window, centered on the simulator vehicle. The basic idea of the moving window is to avoid the simulation of vehicles several miles ahead or behind the simulator vehicle, which is not very efficient. However, the window cannot be too small. Firstly, the size of the window is constrained by the sight distance. The window must at least be as wide as the sight distance, so as to avoid vehicles "popping up" in front of the simulator vehicle. Secondly, the window must be large enough for the traffic to be realistic and also to allow for speed changes of the simulator vehicle.

In order to get a wide enough window but at the same time limit the computational efforts, the window is divided into one inner and two outer regions. It is important that the vehicles in the closest neighborhood of the simulator vehicle behaved like real drivers. Vehicles traveling in the inner region are therefore simulated according to sub-models for car-following, overtaking and speed adaptation, etc. The inner area is called the simulated area. The behavior of vehicles traveling further away from the simulator vehicle is less important, and these vehicles, traveling in the outer regions, are simulated according to a simple mesoscopic model. When approaching the simulated area, these vehicles become candidates to move into the simulated area. The outer regions are therefore referred to as candidate areas. At the limits of the candidate areas, vehicles traveling out of the system are removed from the model and new vehicles are generated.

3.2.2   Generation of new vehicles

Vehicles traveling much slower or faster than the simulator vehicle travel out of the simulated area, into the candidate areas and finally out of the system. Consequently, the system would become empty if no new vehicles are generated. Since the model did not include intersections or ramps, all new vehicles are generated at the edges of the window. Since the edges moved with the speed of the simulator vehicle, new vehicles could not be generated in the same way as in ordinary traffic simulation models, where new vehicles are generated at the geographical places defining an origin in the simulated network. The idea are instead to only generate faster vehicles at the edge behind and slower vehicles at the edge in front of the simulator vehicle. In ordinary traffic simulation models, the vehicle arrival time is drawn from a time headway distribution. The average time headway between arriving vehicles is calculated as the inverse of the traffic flow. If the arrival time between faster vehicles generated behind the simulator vehicle are calculated in this way, the average distance between them would be equal to the average distance between vehicles, which would be incorrect and lead to a traffic composition differing from the one specified. In order to deal with this problem and to ensure that the specified traffic conditions are achieved, a new generation algorithm is developed. The basic principle of the developed generation model is to generate new temporary vehicles until a faster/slower one is generated after which the temporary ones are discarded. The time headways and desired speeds of the discarded vehicles are then used to calculate a correct arrival time for the faster/slower vehicle according to

$$\Delta T = \frac{\sum_{i=1}^{n} (\Delta t_i \cdot v_i)}{v_n - v_{\mathrm{DS}}} \tag{3.1}$$

where Δt_i and v_i are the time headway and speed of the i:th discarded vehicle, respectively, v_n is the speed of the accepted vehicle and v_{DS} is the current speed of the driving simulator vehicle. On rural roads, the time headway Δt_i is drawn from an exponential distribution, whereas on freeways, the time headway is instead assumed to follow the time headway distribution developed in the HUTSIM/TPMA model (Blad, 2002), specially developed for time headways on Swedish freeways. The generation model is designed and tested for generation of traffic in non-saturated conditions. For saturated conditions, it may be necessary to use other types of time headway distributions.

In order to limit the computational effort and to avoid the algorithm getting "stuck" trying to generate faster vehicles when the simulator vehicle is driving very fast, new vehicles are only generated behind the simulator vehicle when it is traveling slower than the highest speed in the current desired speed distribution, and analogously for the edge in front of the simulator vehicles. For the same reason, the number of tries at each time step is restricted, currently to 10 presumptive new veh/time step, i.e. $n \leq 10$.

In order to avoid too long arrival times, the speed of the generated vehicle, v_n, has to differ by at least 5% from the simulator vehicle's speed, v_{DS}. If the speed is within this range, i.e. $v_{DS} < v_n \leq 1.05 \cdot v_{DS}$ for the edge behind the simulator vehicle and $0.95 \cdot v_{DS} < v_n \leq v_{DS}$ for the edge in front, respective speeds of $1.05 \cdot v_{DS}$ and $0.95 \cdot v_{DS}$ can instead be used in the calculations of the arrival time.

Vehicles in the oncoming direction on rural roads are generated according to the vehicle platoon generation model presented by Brodin and Carlsson (1986). For oncoming vehicles on freeways, the HUTSIM/TPMA model is employed (Blad, 2002).

3.2.3   Sub-models for driving behavior and vehicle movement

Car-following
The utilized car-following model is based on the HUTSIM/TPMA model (Kosonen, 1999) with some modifications, see Janson Olstam (2005) for details. The car-following model employs three regimes: *Free*, *Stable*, and *Forbidden*, and these regimes are defined by headways. The forbidden headway $d_f(v_n, v_{n-1})$ is a function of the speed of the follower and the leader. The forbidden headway also depends on a driver-dependent minimum desired time gap, an average normal deceleration rate and a minimum distance between stationary vehicles. The stable regime is defined as the regime enclosed by the forbidden regime and the free regime. The width of the stable regime, W_{stable}, depends on $d_f(v_n, v_{n-1})$.

When a vehicle is in the free regime, $x_{n-1} - x_n > d_f + W_{stable}$, the driver accelerates or decelerates in order to reach the desired speed. In the stable regime, $d_f < x_{n-1} - x_n \leq d_f + W_{stable}$, the driver refrains from taking any action. If a vehicle enters the forbidden regime, $x_{n-1} - x_n \leq d_f$, the driver decelerates in order to reenter the stable regime.

For free accelerations, the acceleration model presented in Brodin and Carlsson (1986) is used, in which the acceleration for vehicle n could be calculated as

$$a_n = \frac{p_n}{v_n} - (C_A)_n \cdot v_n^2 - (C_{R_1})_n - (C_{R_2})_n \cdot v_n - g \cdot i(x_n) \tag{3.2}$$

where p_n is the power/mass ratio for vehicle n, C_A, C_{R_1}, and C_{R_2} are vehicle-type-dependent air and rolling resistance coefficients, and g is the gravitational acceleration constant. The function $i(x_n)$ represents the road incline at the position x_n of vehicle n.

Lane-changing
The lane-changing model is also based on the HUTSIM/TPMA model (Kosonen, 1999), with some minor modifications, see Janson Olstam (2005) for details. In this model, a pressure function is used to decide whether or not a driver desired to change lanes. The pressure is defined as

$$P = \frac{(v_{des} - v_{obs})^2}{2 \cdot s} \tag{3.3}$$

where v_{des} is the desired speed of the rearmost vehicle, v_{obs} is the speed of the obstacle vehicle, and s is the distance between the two vehicles. The decision to change to the

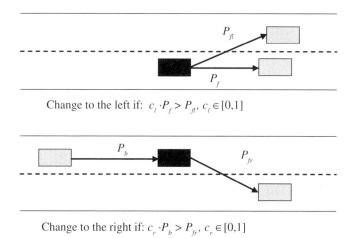

Change to the left if: $c_l \cdot P_f > P_{fl}$, $c_l \in [0,1]$

Change to the right if: $c_r \cdot P_b > P_{fr}$, $c_r \in [0,1]$

Fig. 3.1 The lane-changing logic, based on the model presented by Kosonen (1999).

left is based on the pressure to the closest vehicle in front in the own lane and to the first vehicle in the left lane according to the rules presented in Figure 3.1. For lane changes to the right, the pressure from the back vehicle in the left lane is used. The parameters c_l and c_r are calibration parameters that control the willingness to change lanes to the left and right, respectively.

Speed adaptation
The sub-model for determining a vehicle's desired speed at a section is based on the speed adaptation model used in VTISim (Brodin and Carlsson, 1986). This model describes speed adaptation on rural roads and has therefore been recalibrated for freeways, see Janson Olstam (2005). The model starts from an average basic desired speed, v_0, and this speed is then reduced with respect to the speed limit, road width, and curvature to an average desired speed, v_3, for a specific road section. The desired speed for a vehicle n at a road section is finally calculated as

$$v_{3n} = \left(v_{0n}^Q - (1-\alpha) \cdot \left(v_0^Q - v_3^Q \right) \right)^{1/Q} \tag{3.4}$$

where v_{0n} is the basic desired speed of vehicle n and $0 \le \alpha \le 1$ is a vehicle–type-dependent parameter, equal to 0 for cars. The parameter Q is a transformation measure that depends on the reason for reduction. $Q = 1$ implies a parallel shift of the basic desired speed distribution curve. Values of $Q < 1$ also imply an anticlockwise rotation of the distribution curve around the median, which leads to the desired speed of drivers with high basic desired speeds being more affected than drivers with low basic desired speeds.

Overtaking
The model used for overtaking on rural roads with oncoming traffic is based on the VTISim model (Brodin and Carlsson, 1986). This model takes into account probability

functions that describe the probability of starting an overtaking as a function of the available overtaking gap. The probability function is defined as

$$P\left(d_{\text{gap}}\right) = e^{-A \cdot e^{-k \cdot d_{\text{gap}}}} \tag{3.5}$$

where d_{gap} is the available gap, that is min{distance to oncoming vehicle, distance to natural sight obstruction}, and A and k are constants that depend on: the type of overtaking {flying, accelerated}, the type of sight limitation {oncoming vehicle, natural}, the type and speed of the vehicle being overtaken, as well as the current road width. A flying overtaking is an overtaking that starts when the driver has caught up with a slower vehicle but not yet started to adjust its speed to that vehicle. If the driver cannot execute a flying overtaking it has to decelerate and adjust its speed to the preceding vehicle and may later try to execute an accelerated overtaking. Calibrated values of A and k for Swedish road conditions are available in Carlsson (1993).

All drivers are assumed to have a higher desired speed during overtakings, currently set to a temporary increase of 10 km/h. Car drivers are also assumed to use higher acceleration rates during overtakings, which is modeled as an increase in their power/mass ratio.

When overtaking, the overtaking vehicle must continuously reevaluate the distance to the vehicle in the oncoming lane and the remaining distance of the overtaking. These parameters were not included in the model presented in Brodin and Carlsson (1986), and thus the overtaking model is enhanced with a more detailed modeling of overtakings including decision rules for aborting an overtaking. The developed model for overtaking abortions is based on the assumption that a driver takes action if the time to collision, TTC, with the oncoming vehicle is less than the estimated time left of the overtaking, or in other words, if $\text{TTC} + t_{\text{safety}} < t_{\text{left}}$, where t_{safety} is a safety margin and t_{left} is the time left of the overtaking estimated as

$$t_{\text{left}} = -\frac{v_n - v_{n-1}}{a_n} + \sqrt{\left(\frac{v_n - v_{n-1}}{a_n}\right)^2 + 2\frac{\Delta d}{a_n}} + 0.5 \cdot t_{\text{change}} \tag{3.6}$$

here, $\Delta d = x_{n-1} - x_n + l_n + d_{\text{min}}$, in which d_{min} is the critical lag gap for lane changes to the right and t_{change} is the time it takes to perform the lane change back to the normal lane. The variables v_n, x_n, v_{n-1}, and x_{n-1} refer to the speed and position of the overtaking vehicle and the vehicle being overtaken, respectively. The acceleration a_n is calculated according to Eq. (3.2). In situations where $\text{TTC} + t_{\text{safety}} < t_{\text{left}}$ and the driver has not yet passed the lead vehicle, the driver is assumed to abort the overtaking. The driver then falls back and merges into the normal lane behind the lead vehicle. If the vehicle is side-by-side or has passed the lead vehicle, the driver will instead increase the desired speed to a level required to end the overtaking without colliding with the oncoming vehicle. If the vehicle's power/mass ratio is too low in order to be able to accelerate to the new desired speed, verified via Eq. (3.2), the vehicle is temporarily assigned a new power/mass value. However, if the power/mass ratio required to drive at the new desired speed exceeds the maximum power/mass ratio for the current vehicle type, the driver aborts the overtaking and falls back in order to merge into the normal lane behind the overtaken vehicle.

Lateral movements

The vehicle's lateral position is assumed to change only as a result of a lane-change. When driving in a lane, the vehicles are assumed to drive in the middle of it. This simplification does so far not seem to affect simulator drivers' experienced realism of the surrounding vehicles. However, for a more realistic modeling, some kind of lateral disturbance factor or function should be added. During a change of lane, two approaches for modeling the lateral movements have been tested. In the first, the lane-changing movements of the vehicles are assumed to follow a sine-curve. In the second alternative, the movements follow a function using a second grade polynomial in the beginning and end of the movement, and a linear relationship in between. Both approaches appeared quite realistic on freeways, where lane-changing movements are carried out during quite long times, i.e. about 4–6 seconds according to measurements presented in Liu and Salvucci (2002). However, on rural roads "lane-changing" movements are sometimes executed during much shorter times, for instance at evasive maneuvers or when aborting an overtaking. It seems that none of the two functions offered a realistic representation of lateral movements during fast lane-changes. Another drawback is that both functions assumed that all started lane-changes are completed. The functions cannot model the lateral movements when a driver decides to abort an ongoing lane-change. In order to overcome these drawbacks, a more advanced steering model is required, perhaps a model similar to the one presented by Salvucci et al. (2001) or one based on control theory.

Brake lights and turning signals

In standard traffic simulation applications, there is no need for simulating occurrences like the use of turn signals or brake lights since all vehicle actions are known within the model. However, when simulating traffic for a driving simulator, it is important to model both turn signals and brake lights, otherwise such signals will not be visible for the simulator driver. Brake lights have in this work been presumed to be on for deceleration rates higher than that of an engine, taken as 0.5 m/s^2 (in reality the engine deceleration rate depends on the vehicle's speed and gear). Drivers are assumed to use the turn signals with a certain probability that differs between lane-changes to the right and left as well as between freeways and rural roads. For instance, when driving on freeways, drivers are assumed to use the left turn signal more often than the right one.

3.2.4   The candidate areas

In the candidate areas the vehicles are simulated by a new mesoscopic traffic simulation model. The traffic stream is still represented by individual vehicles but interactions are modeled at an aggregated level using speed-flow functions from SRA (2001). The speed of vehicle n is calculated as

$$v_n = \left(f(q)^Q + \left(\left(v_n^{\text{des}} \right)^Q - f(0)^Q \right) \right)^{1/Q} \tag{3.7}$$

where v_n^{des} is the desired speed of vehicle n, q is the traffic flow, and $f(q)$ is the average travel speed at a traffic flow of q veh/h. The parameter Q controls the rotation of the speed distribution curve. Values of $Q < 1$ imply that the speed of vehicles traveling fast will decrease more than the speed of vehicles driving slowly. Two different values for Q have been tested: 1 and -0.2. $Q = 1$ implies no rotation, and vehicles are thus able to drive as much faster or slower than the average speed as they would do during free flow conditions. $Q = -0.2$ is the value used for speed adaptation to speed limits in the similar speed adaptation model presented by Brodin and Carlsson (1986). The model seemed to perform well at both values of Q, but a value of $Q < 1$ appeared to be more realistic since the speed variation between drivers normally decreases with an increasing traffic flow under non-saturated conditions. Further calibration and evaluation is needed before a recommendation can be made.

Apart from the reduction of speed according to Eq. (3.7), the candidate vehicles travel unconstrained with regard to the surrounding traffic. When a candidate vehicle catches up with another candidate vehicle it can always overtake the preceding vehicle without any loss in time.

A candidate vehicle that reaches either boundary of the simulated area is only allowed to travel into the simulated area if there is a sufficient distance to the first vehicle in this area. For a vehicle desiring to enter the simulated area from the candidate area behind the simulator vehicle, the car-following model is used to deduce whether it can do so or not. The vehicle is allowed to enter the simulated area if it can do so without decelerating, thus when the car-following model returns a non-negative acceleration. If this is not the case, the vehicle adopts the acceleration given by the car-following model and is placed at the edge between the candidate area and the simulated area. The vehicle gets a new opportunity to pass into the simulated area in the next time step. While waiting for a sufficient gap, the candidate vehicle adjusts its speed in order to avoid strong decelerations when entering the simulated area. In the freeway environment, cars are also given the possibility of entering the simulated area in the left lane. Whether this is possible or not is verified using both the lane-changing and the car-following model. For vehicles in the candidate area in front of the simulator vehicle, a similar but somewhat different approach is employed. The simulated vehicle closest to the candidate area treats the first vehicle in the candidate area as any other simulated vehicle. It thus uses the car-following model to adjust the speed and the lane-changing or overtaking model in order to decide whether it should try to overtake the candidate vehicle or not.

3.3   VALIDATION

The primary outputs of the developed model are made up of the behavior of the simulated vehicles. The primary output is thus at a microscopic level and not at a macroscopic level, which is the case for most applications of traffic simulation. The simulation model has therefore been validated at a microscopic level. Validation at a macroscopic level is still of interest and should be conducted in future studies. The overall objective for the present application is that the simulator drivers consider the behavior of the surrounding vehicles as realistic. If this is not the case, the driver of the simulator vehicle may behave differently as compared to when driving a real car. The problem is that "realistic" is difficult to define. It is also hard to state how realistic the behavior has to be in order for the model to be valid. The ultimate goal is, of course, to obtain a model where the simulator driver cannot conclude whether a surrounding vehicle is driven by another human or by a computer.

Two separate validation studies were performed within the scope of this project. The first was a study on overtaking rates of the driving simulator vehicle and the second a study on how realistically human drivers consider the simulated drivers to behave.

3.3.1 Overtaking rates

An important part connected to the observed realism is the number of vehicles that catches up with the driving simulator and the number of vehicles that the simulator driver catches up with. When driving at a certain speed, it may be difficult to determine whether the number of vehicles that overtakes you is comparable to when driving on a real road. However, you would certainly react if the proportion between vehicles that catch up with you (passive catch-ups) and those that you catch up with (active catch-ups) is unrealistic. Within the project, active and passive catch-ups generated by the model have been compared with an analytical expression for estimating the number of catch-ups of a floating car, originally presented by Carlsson (1995). Passive catch-ups can, according to this expression, be estimated as

$$U_p = qL \int_{v_0}^{\infty} \left(\frac{1}{v_0} - \frac{1}{v} \right) f_t(v)dv \qquad (3.8)$$

where

  q : is the traffic flow (veh/h),
  L : is the length of the observed road section,
  v_0 : is the speed of the studied vehicle, and
  $f_t(v)$: is the time mean speed distribution.

The number of active catch-ups is calculated in a similar way. One of the underlying assumptions for these functions is that all vehicles can overtake one another without any time delay. The equations can thus be expected to give upper limits on the number of active and passive catch-ups.

The values from the model were generated by not only simulating the surrounding vehicles but also by simulating the driving simulator vehicle. Simulations were performed at various desired speeds of the simulator vehicle and at varying traffic flows. For rural roads, the simulated values corresponded quite well to the analytical calculation, as can be seen in the example with 400 veh/h in each direction in Figure 3.2. However, the simulated number of active catch-ups on freeways appeared too low, as demonstrated in the example with 1000 veh/h in Figure 3.3. The reason for this was believed to be a too high frequency of lane changes. It was however noticeable that the analytical expression constituted an upper limit and that the simulated values were generally smaller than their corresponding analytical counterparts. The high frequency of lane changes seemed to be due to a lack of anticipation in the lane-changing model. These deficiencies in the lane-changing model resulted in the relationship between average speed and traffic flow in the simulated area (the

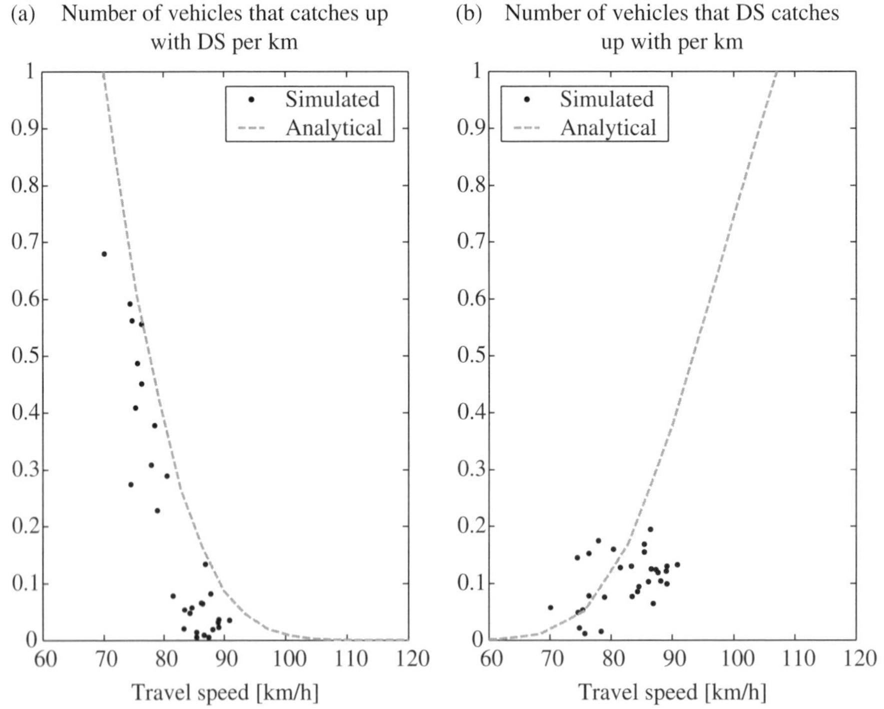

(a)   Number of vehicles that catches up
       with DS per km

(b)   Number of vehicles that DS catches
       up with per km

Travel speed [km/h]

Travel speed [km/h]

Fig. 3.2 The simulated and calculated number of (a) passive and (b) active catch-ups per km of the driving simulator vehicle (DS) on a straight and plain rural highway with oncoming traffic.

micro-simulated area) differing from the one used in the mesoscopic model and the traffic generation model. Consequently, the simulator driver would not be able to drive as fast as expected by the generation and the mesoscopic models, which would result in the vehicles in the candidate areas generally driving faster than the vehicles in the simulated area. This would give rise to fewer active catch-ups.

3.3.2   User evaluation

It has in the present work also been attempted to validate the simulation model by using human drivers as measurement devices. The human mind is a very useful tool that can be employed both for detecting unrealistic behavior and for obtaining statements on how realistic the behavior of the simulated vehicles was. In order to get these statements, a small driving simulator experiment with ten participants, three females and seven males, was conducted. The experiment was performed in the VTI Driving Simulator III (VTI, 2006). After 10 minutes of warm-up driving, the participants drove 15 minutes along a rural road and 15 minutes on a freeway. For the rural scenario, part of the two-lane highway Rv 34 between Målilla and Hultsfred was used. This road is 9 meters wide with a carriageway of 7 meters. The posted speed limit is 90 km/h. For the freeway scenario, part of the European Road E4 between Linköping and Norrköping was utilized. This road has two lanes in each direction and

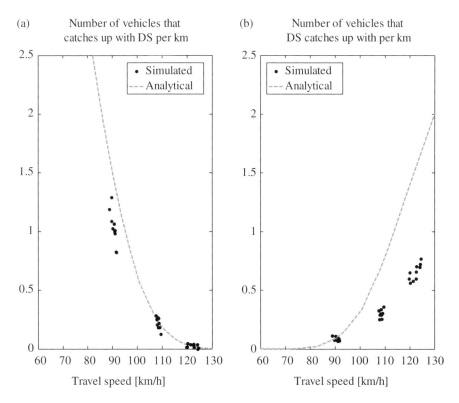

Fig. 3.3 The simulated and calculated number of (a) passive and (b) active catch-ups per km of the driving simulator vehicle (DS) on a straight and plain freeway.

a posted speed limit of 110 km/h. All intersections and ramps were removed from the scenario roads. The Rv 34 and E4 were chosen since they are good representatives of Swedish two-lane highways and freeways. The participants were not exposed to any critical situations and were instructed to drive as they normally do in similar road environments.

After the ride, the participants were asked to comment on the simulated vehicles' behavior by filling out an evaluation form and answering various interview questions. The participants were first asked about the realism of other parts of the driving simulator (steering, accelerator pedal, screen depth, etc) in order to avoid that opinions of these matters influencing the answers on the forthcoming questions. Subsequently, the participants were asked to estimate to what extent they thought that the other road users behaved like real drivers. The results were quite good. On a scale from 1 to 7, the following scores were received: [5,7,7,5,3,7,6,6,3,5] and [5,7,5,5,7,7,4,6,5,3] for the two-lane highway and the freeway, respectively. The most common comments were that the simulated drivers drove aggressively on the rural road and that they drove slower than in reality. Some participants also thought that some of the simulated drivers remained in the left lane on the freeway for a long time before changing back to the right lane. Other comments given by some of the participants were: "It felt OK" and "This is what it looks like in dense traffic".

The reason for the aggressive behavior was probably a too small safety margin during overtakings. Later tests with a larger safety margin indicated that this solved the problem. A probable reason for certain of the participants thinking that the simulated drivers sometimes started overtakings at risky places, e.g. places with limited sight, was that it can be quite hard for the participants to distinguish objects far away on the simulator screen while the simulated vehicles had "perfect" vision. The best way to solve this would probably be to limit the simulated vehicles' sight in order for it to better correspond to the sight distances experienced by the simulator driver.

The fact that the simulated drivers seemed to drive slowly could be explained by the speedometer in the driving simulator showing the actual speed and not a speed that is 5–8 km/h higher, which is the case in most real cars. This resulted in the participants experiencing a certain speed in the simulator to be faster than they normally experienced it, thus giving the impression that the surrounding vehicles drove slower than in real life. This can be easily fixed by adjusting the speed in the speedometer to better correspond to real conditions.

3.4   CONCLUDING REMARKS AND FUTURE RESEARCH

The simulation model presented in this chapter was able to generate and simulate surrounding traffic for a driving simulator, on rural roads and on freeways. The model generated realistic streams of vehicles both in the same and the oncoming direction as that of the simulator vehicle. The model included, among other things, a new technique for generating traffic on a moving area around a specific vehicle, an enhanced version of the VTISim (Brodin and Carlsson, 1986) overtaking model, and a mesoscopic simulation model for road links. The model was integrated and tested within the VTI Driving simulator III, and the performed validation showed that the simulated vehicle-driver units behaved – to a large extent – as real drivers. The results from the validation studies also indicated that further enhancements and calibration of for example the sub-models for overtaking and lane-changing were required.

Not only would the development of a model for the generation and simulation of surrounding traffic for driving simulators increase the realism in driving simulators. It would also create possibilities to develop new or enhance existing traffic simulation models. Data concerning all movements, including the driving simulator vehicle's movements, can be gathered. This data can then be used to study, for example, car-following, lane-changing, and overtaking behaviors in order to create more realistic sub-models for driving behavior. The combination of a driving simulator and a traffic simulation model also creates new additional means for validating traffic simulation models. The validity of a model can now also be verified by driving around in the simulated traffic, and such subjective or qualitative analysis can be a good complement to the traditional comparisons of speeds, flows, queue lengths, etc.

The presented model was only able to simulate road links, i.e. roads without intersections and ramps. In order to be really useful, the model must also include modeling of on- and off-ramps on freeways. This implies a detailed modeling of lane-changing and acceleration behavior in merging situations. One problem here can be that some merging models use priority rules such as "closest to the merging point goes first". Such approaches cannot be employed in this kind of application since driving simulator drivers may not follow this behavior. To achieve a more complete modeling of rural roads, the model has to be extended to include modeling of intersections and roads with a barrier between oncoming lanes, e.g., so called 1 + 1 and 2 + 1 roads.

The developed model can only be used in driving simulator scenarios in which participants would not be exposed to critical events. However, many driving simulators scenarios include critical events and an important research need thus includes to be able to create scenarios that combine autonomous simulations of vehicles and critical events, see e.g., Janson Olstam and Espié (2007). The basic idea would be to use the autonomous simulation model to simulate the vehicles during the time between the predetermined critical situations. When approaching the point in time or space where the critical event is going to take place, the simulation of the surrounding vehicles should, in a way that is unnoticeable to the driver, turn from autonomous mode to becoming totally controlled according to the defined scenario.

3.5   ACKNOWLEDGMENTS

Thanks are due to the Swedish National Road Administration (SNRA) for funding this work, to Jan Lundgren (ITN/LiTH) for invaluable comments and support, to Mikael Adlers (VTI) for his priceless help during the integration with the VTI driving simulator. Many thanks are also expressed to the colleagues at VTI for sharing their expertise within the driving simulator and traffic simulation area.

3.6   REFERENCES

Ahmad, O., and Papelis, Y. (2001) "A comprehensive microscopic autonomous driver model for use in high-fidelity driving simulation environments", in Proceedings of the 81st Annual Meeting of the Transportation Research Board, Washington D.C., USA.

Bang, B., and Moen, T. (2004) *Integrating a Driving Simulator and A Microscopic Traffic Simulation Model*, Trondheim, Norway, SINTEF (unpublished paper).

Barceló, J., and Casas, J. (2002) "Dynamic network simulation with AIMSUN", in Proceedings of the International Symposium on Transport Simulation, Yokohama, Japan. http://www.aimsun.com/Yokohama_revised.pdf.

Bayarri, S., Fernandez, M., and Perez, M. (1996) "Virtual reality for Driving Simulation", *Communications of the ACM*, 39(5):72–76.

Blad, P. (2002) "Part E: Traffic Generators", in TPMA – Model-1 Final Report, F. Davidsson, I. Kosonen and A. Gutowski. Stockholm, Sweden, Centre for Traffic Research (CTR). http://www.infra.kth.se/ctr/projekt/tpma/tpma_en.htm.

Bonakdarian, E., Cremer, J., Kearney, J., and Willemsen, P. (1998) "Generation of ambient traffic for real-time driving simulation", in Proceedings of IMAGE Conference, Scottsdale, Arizona, USA.

Brodin, A., and Carlsson, A. (1986) *The VTI Traffic Simulation Model – A Description of the Model and Programme System*. VTI Meddelande 321A, Swedish National Road and Transport Research Institute (VTI), Linköping, Sweden.

Carlsson, A. (1993) Beskrivning av VTIs trafiksimuleringsmodell (Description of VTIs traffic simulation model, In Swedish), VTI Notat T 138, Swedish National Road and Transport Research Institute (VTI), Linköping.

Carlsson, A. (1995) Omkörning av floating car (Number of overtakings of a floating car, in Swedish). Linköping, Sweden, VTI (unpublished working paper).

Champion, A., Mandiau, R., Kolski, C., Heidet, A., and Kemeny, A. (1999) "Traffic generation with the SCANeRII simulator: towards a multi-agent architecture", in Proceedings of the Driving Simulator Conference, DSC'99, Paris, France.

El Hadouaj, S., and Espié, S. (2002) "A generic road traffic simulation model", in Proceedings of the ICTTS (Traffic and transportation studies), Guilin, China.

Espié, S. (1995) "ARCHISIM: Multiactor parallel architecture for traffic simulation", in Proceedings of the second world congress on Intelligent Transport Systems '95, Yokohama, Japan.

Janson Olstam, J. (2005) A Model for Simulation and Generation of Surrounding Vehicles in Driving Simulators. Licentiate Thesis at Institute of Technology, Linköpings universitet, Norrköping. LiU-TEK-LIC 2005:58.

Janson Olstam, J., and Espié, S. (2007) "Combination of autonomous and controlled vehicles in driving simulator scenarios", in Proceedings of the Road Safety and Simulation (RSS2007), Rome, Italy.

Jenkins, J. M. (2004) Modeling the Interaction Between Passenger Cars and Trucks. Doctorial Thesis at Texas A&M University, Texas, USA.

Kosonen, I. (1999) HUTSIM – Urban Traffic Simulation and Control Model: Principles and Applications. Doctorial Thesis at Department of Civil and Environmental Engineering, Helsinki University of Technology, Helsinki, Finland.

Kuwahara, M., Tanaka, S., Kano, M., Furukawa, M., Honda, K., Maruoka, K., Yamamoto, T., Shiraishi, T., Hanabusa, H., and Webster, N. (2005) "An enhanced traffic simulation system for interactive traffic environment", in Proceedings of the IEEE Intelligent Vehicles Symposium, Las Vegas, USA.

Liu, A., and Salvucci, D. D. (2002) "The time course of a lane change: driver control and eye-movement behavior", Transportation Research, 5(2):123–132.

PTV (2003) VISSIM User Manual – Version 3.70. PTV Planung Transport Verkehr AG, Karlsruhe, Germany.

Salvucci, D. D., Boer, E. R. and Liu, A. (2001) "Toward an integrated model of driver behavior in a cognitive architecture", Transportation Research Record, 1779:9–16.

SRA (2001) "Appendix 1 – VQ-diagrams", in Effektsamband 2000 – Nybyggnad och förbättring (Effects of different road measurements – Construction and improvement, in Swedish). Borlänge, Sweden, Swedish Road Administration, Publikation 2001. 78:336–348.

Wright, S. (2000) Supporting Intelligent Traffic in the Leeds Driving Simulator. Doctorial Thesis at School of Computing, University of Leeds, Leeds.

VTI (2006) Driving Simulators at VTI. http://www.vti.se/templates/Page_3257.aspx, accessed May 15th, 2006.

PART II

Lane Changing

CHAPTER 4

INTEGRATED LANE-CHANGING MODELS

Moshe Ben-Akiva, Charisma Choudhury, Tomer Toledo

This chapter summarizes a series of advances in lane changing models aiming at providing a more complete and integrated representation of drivers' behaviors. These advances include the integration of mandatory and discretionary lane changes in a single framework, the inclusion of an explicit target lane choice in the decision process and the incorporation of various types of lane-changing mechanisms, such as cooperative lane changing and forced merging. In the specifications of these models, heterogeneity in the driver population and correlations among the various decisions a single driver makes across choice dimensions and time are addressed. These model enhancements were implemented in the open source microscopic traffic simulator of MITSIMLab, and their impact was demonstrated in validation case studies where their performance was compared to that of existing models. In all cases, a substantial improvement in simulation capability was observed.

4.1   INTRODUCTION

Lane changing has a significant impact on traffic flow. Lane-changing models are therefore an important component in microscopic traffic simulators, which are becoming the tool of choice for a wide range of traffic-related applications at the operational level. A number of lane-changing models have been proposed and implemented in various simulators in recent years (see Toledo, 2006 for a review). While their details vary, the general structure of these models displayed in Figure 4.1 is similar. Most models classify lane changes as either mandatory or discretionary (e.g., Ahmed, 1999; Ahmed et al., 1996; Gipps, 1986; Halati et al., 1997; Hidas, 2002; Hidas and Behbahanizadeh, 1999; Yang and Koutsopoulos, 1996; Zhang et al., 1998). A mandatory lane change (MLC) occurs when a driver must change lanes to follow a path. When an MLC is required, it overrides any other considerations. A discretionary lane change (DLC), on the other hand, takes place when a driver changes to a lane perceived to offer better traffic conditions. Gap acceptance models are used to model the execution of lane changes. The available gaps are compared to the smallest acceptable gap (critical gap)

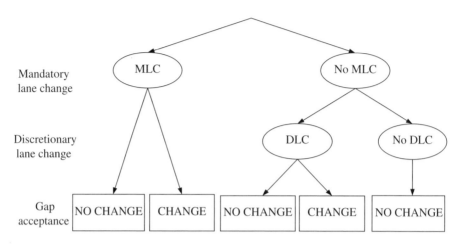

Fig. 4.1 The generic structure of lane-changing models.

and a lane-change is executed if the available gaps are greater. Gaps may be defined either in terms of time or free space. Most models also make a distinction between the lead gap (i.e., the gap between the subject vehicle and the vehicle ahead of it in the lane it is changing to) and the lag gap (i.e., the gap between the subject vehicle and the vehicle behind it in the lane it is changing to) and require both to be acceptable.

The present chapter summarizes several enhancements that have been made to this generic lane-changing model in order to improve its realism and address several limitations. The organization of the chapter is as follows: first, the general methodology that was used to develop each one of the models presented in this chapter is given. The next three sections present enhancements that have been made to the lane-changing model: integration of mandatory and discretionary lane changes in a single framework, explicit modeling of the choice of the target lane and a model that incorporates courtesy behavior and forced merging in the lane-changing process. For each of these models, the modified structure of the lane-changing decision process is presented and the limitations of the basic model that it addresses are discussed and demonstrated with a real-world case study. The specification of all these models account for the heterogeneity in the behavior population and for correlations among the decisions a single driver makes over choice dimensions and time. The mathematical formulation that permits the capturing of these effects is presented in the next section. The final section concludes our findings and discusses directions for further enhancements.

4.2   METHODOLOGY

All the models that are presented here were developed using the process shown in Figure 4.2, which involves both disaggregate and aggregate data. Disaggregate data, consisting of detailed vehicle trajectories at a high time resolution, are used in the model estimation phase, in which the model is specified and explanatory variables, such as speeds and relations between the subject vehicle and other vehicles, are generated

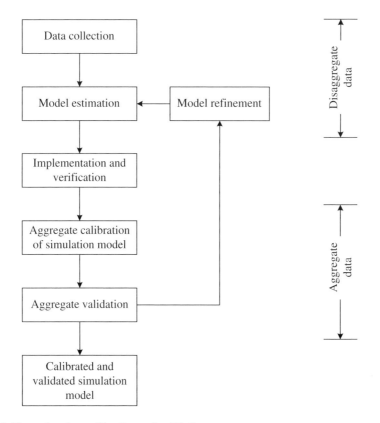

Fig. 4.2 The estimation, calibration and validation process.

from the vehicle coordinates extracted from the trajectory data. The model parameters are estimated with a maximum likelihood technique in order to match observed lane changes having occurred in the trajectory data. This estimation approach does not involve the use of a traffic simulator, and so the estimated models are independent of simulators.

In order to demonstrate the benefits that may be derived from using the modified models, they must be validated and demonstrated within a microscopic traffic simulator incorporating not only the lane-changing models under investigation, but also other driving behavior models, such as acceleration models. As a result, the estimated model needs to be implemented within a microscopic traffic simulator. MITSIMLab (Yang and Koutsopoulos, 1996) was employed in all the cases described herein. In the validation case studies, aggregate data, which is significantly cheaper to collect and in many cases readily available, could be used. Part of the aggregate dataset was first utilized to adjust key parameters in the lane-changing model as well as parameters of other behavior models, and to estimate the travel demand on the case study network. This aggregate calibration problem was formulated as an optimization problem, seeking to minimize a function of the deviation of the simulated traffic measurements from the observed measurements and of the deviation of calibrated values from their a-priori estimates, when available (Toledo et al., 2003). The rest of the data was used for the validation itself, which was based on a comparison of measures of performances that could be calculated from the available data with corresponding values from the simulator, e.g., sensor speeds and flows, the distribution of vehicles among the lanes, the amount and locations of lane changes. The calibration and validation methodology is outlined in Figure 4.2 (detailed in Toledo and Koutsopoulos, 2004).

4.3   INTEGRATION OF MLC AND DLC

As noted above, most models classify lane changes as either mandatory or discretionary, with the former overriding the latter. This separation implies that there are no trade-offs between mandatory and discretionary considerations. For example, a vehicle on a freeway which intends to take an off-ramp will not overtake a slower vehicle if the remaining distance to the off-ramp is below a certain threshold value, regardless of the speed of that vehicle. Furthermore, in order to implement the MLC and DLC models separately, rules that dictate when drivers begin to respond to MLC conditions need to be defined. However, this point is unobservable, and so only judgment-based heuristic rules, which are often defined by the remaining distance from the point where the MLC must be completed, are employed.

The model shown in Figure 4.3 integrates MLC and DLC into a single utility model. Variables that capture the need to be in the correct lane and to avoid obstacles as well as variables that capture the relative speed advantages and ease of driving in the current lane, and in the lanes to the right and to the left, are all incorporated in a single utility model that takes into account the trade-offs among these variables. An important goal that affects drivers' lane-changing behavior in this model is following the travel path. This goal is accounted for by a group of variables that capture the remaining distance to the point where drivers have to be in specific lanes and the number of lane changes that are needed in order to be in these lanes. Figure 4.4 demonstrates the impact of these variables on the probability that a driver intending to exit a freeway through an off-ramp would target a change to the right. This probability increases when the distance to the off-ramp is smaller (approaching 1 when the distance approaches zero) and when the number of lane changes required increases. Note that with separate MLC and DLC models, the corresponding graph would be a

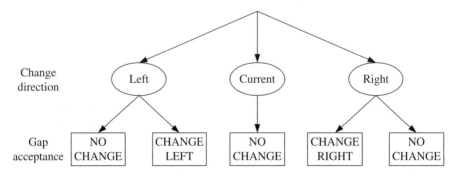

Fig. 4.3 The structure of the integrated MLC and DLC model.

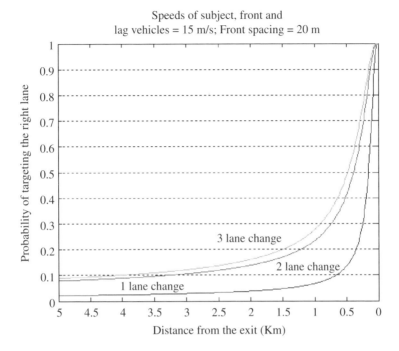

Fig. 4.4 The impact of the path-plan on the probability of targeting the right lane.

step function, with probability 0 when the remaining distance to the ramp is larger than a certain threshold value, and 1 when the distance is smaller.

4.4   EXPLICIT TARGET LANE CHOICE

The decision to seek a lane change and the direction of change in the models introduced so far have been based on an evaluation of the current lane and the adjacent lanes to the right and to the left. Therefore, in these models, the set of lanes that the driver chooses from depends on the lane that the vehicle is currently in. In multi-lane road facilities, only a subset of the available lanes is evaluated. This approach may result in unrealistic behavior in cases where drivers change lanes not because the lane they are changing to is preferable, but as a step on their way to another lane further away in the lane change direction. This type of situation may arise, for example, in multi-lane freeways with dedicated lanes (e.g. HOV lanes). Drivers may change lanes in the direction of the dedicated lane, even to lanes with undesirable characteristics (e.g. slower speeds) in order to eventually enter the dedicated lane, which may provide higher levels of service.

In order to tackle this problem, the model shown in Figure 4.5 (for a driver currently in the second lane from the right, lane 2, of a four-lane freeway) has been tested. This model introduces an explicit target lane selection. Rather than choosing a direction change, drivers choose a target lane among all the available lanes. The target

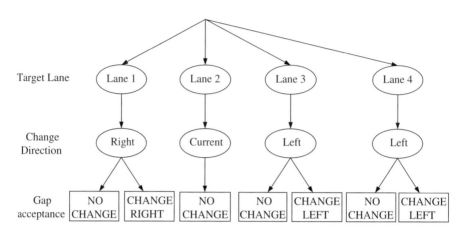

Fig. 4.5 The structure of the model with an explicit target lane choice (current lane is lane 2).

lane is the lane that is perceived as the best lane to be in when multiple factors and goals are taken into account. The direction of a desired lane change, if any, is dictated by the direction of the target lane from the lane that the vehicle is currently in. As with previous models, the completion of the lane change depends on its feasibility, which is captured by gap acceptance models. An estimation of this model with trajectory data demonstrated that important factors affecting the utilities of the various lanes include the microscopic and macroscopic traffic flow characteristics in the lane (e.g., the presence of heavy vehicles, the average speed and density), the impact of the path-plan (e.g., whether it would be a correct lane in order to follow the path), an inertia factor (e.g., whether it is the current lane and if not, the number of lane changes that would be required to reach it) and characteristics of the driver (e.g., aggressiveness).

To demonstrate its usefulness, the model was tested on a section of I-80 in Emeryville, California. This section, shown schematically in Figure 4.6, is six lanes wide, and the left-most lane is an HOV lane that can be accessed at any point in the section. Traffic speeds are significantly higher on this lane as compared to on the other lanes that experience significant queuing and delays during the peak period.

Figure 4.7 shows a comparison of the distribution of vehicles among lanes observed in this section to the ones predicted by two versions of MITSIMLab: one that implements the model with an explicit target lane choice, and another that implements the model described in the previous section, and which is based on a myopic choice of direction change. Overall, the model with an explicit target lane choice matched the observations better, particularly with respect to the usage of the HOV lane. The change direction model underestimated the usage of the HOV lane, mainly because it was not in the set of lane choices of the drivers entering the section from the on-ramp. Consequently, these drivers did not reach this lane. With the explicit target lane model, drivers also evaluated the HOV lane and some chose to change to this lane. As a result, the utilization of the HOV lane was higher and closer to the real-world observations. Additional validation results have been presented in Choudhury et al. (2007a).

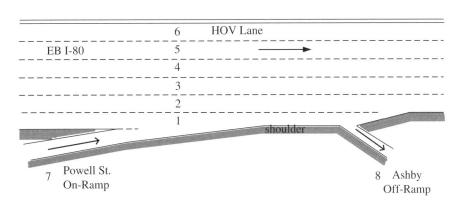

Fig. 4.6 The I-80 site, Emeryville California.

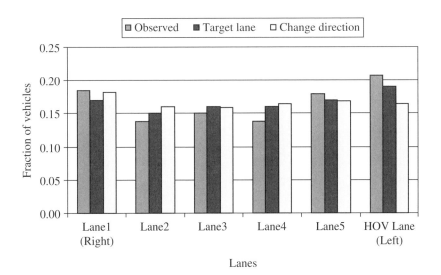

Fig. 4.7 Observed and simulated distributions of vehicles among lanes in the I-80 section.

4.5   COOPERATIVE AND FORCED GAP ACCEPTANCE

The models discussed so far assume that lane changing is executed through gap accept-ance. However, in congested traffic conditions, acceptable gaps may not be available, and so other mechanisms for lane changing are required. For example, drivers may change lanes through courtesy and cooperation of the lag vehicles on the target lane as a result of the latter slowing down to accommodate the lane change. In other cases, certain drivers may become impatient and decide to force their way into the target lane thus compelling the lag vehicle to slow down.

The model shown in Figure 4.8, which was developed for a merging situation, integrates courtesy and forced merging mechanisms with "normal" gap acceptance. The integrated model captures the transitions from one type of merging to the other.

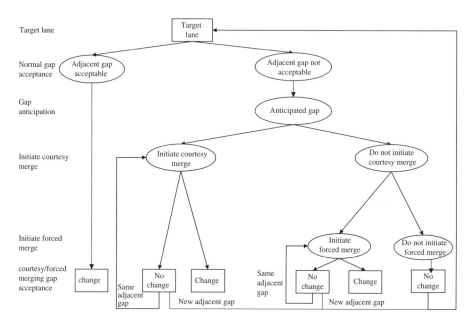

Fig. 4.8 The structure of a merging model that integrates courtesy and forced merging.

In this model the merging driver first evaluates whether or not the available adjacent gap is acceptable. This decision is modeled with standard gap acceptance models that compare the lead and lag gaps with the corresponding critical gaps. If the available gap is acceptable, the driver merges to the mainline. If, on the other hand, the available gap is unacceptable, the driver anticipates what the magnitude of the adjacent gap will be in a short time horizon. The anticipated gap is evaluated based on the magnitude of the available gap and the current speed and acceleration of the lag vehicle. The time horizon over which the driver anticipates the gap may vary across the driver population so as to capture differences in perception and planning abilities among drivers. The anticipated gap reflects the drivers' perception of the courtesy or discourtesy of the lag vehicle. The driver then evaluates whether the anticipated gap is acceptable or not. An acceptable anticipated gap implies that the lag vehicle is providing courtesy to the merging vehicle, and so the driver can initiate a courtesy merge. If the anticipated gap is not acceptable, the lag vehicle is not providing courtesy, in which situation the merging driver may choose whether or not to begin forcing his way into the mainline and compel the lag vehicle to slow down.

A driver that has initiated courtesy yielding or forced merging completes the merge when the available gap is acceptable. Thus, the lane change may not be completed when initiated, but it may rather take more time. However, the model assumes that a driver that has initiated a courtesy or forced merge will continue to use this mechanism until the lane change is complete, or if unsuccessful, until the adjacent gap is no longer available (e.g., having been overtaken by the lag vehicle). Critical gaps for courtesy or forced merging may differ from the ones used in normal lane changing.

Estimation results for this model (Choudhury et al., 2007b) showed that the inclusion of the three merging mechanisms were justified by the data and significantly

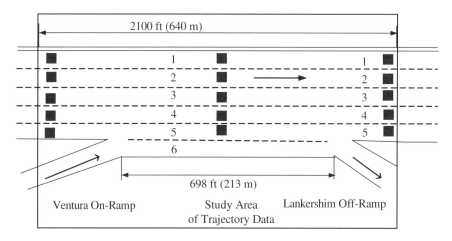

Fig. 4.9 The US101 site, Los Angeles California.

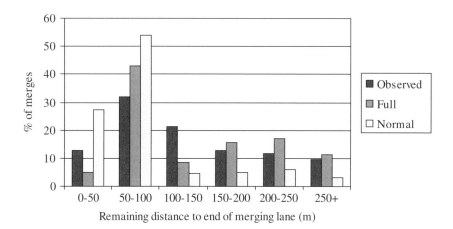

Fig. 4.10 Observed and predicted merge locations.

improved the fit of the model. Critical gaps differed in the various merging mechanisms. In general, the results showed that drivers were willing to accept smaller lead and lag gaps if they perceived that the lag vehicle was courtesy yielding.

   To demonstrate the impact of the inclusion of courtesy and forced merging in the model, a version of MITSIMLab implementing this model was compared with one that only included a standard lane-changing model (Lee, 2006) similar to the one described in the previous section. The network used in the validation was a section of US101 in Los Angeles California, displayed in Figure 4.9. This section gener-ally experiences high congestion during the peak period that was modeled in this case study. Figure 4.10 shows the distribution of locations of merges as a function of the remaining distance to the end of the merging lane, which was observed in the data and predicted by the two MITSIMLab versions. The results indicated that the full model, which incorporated courtesy and forced merging, was able to better

Table 4.1 Comparison of lane-specific flows.

	Normal	Full	Improvement
RMSE (vehicles/5 min)	19.18	13.22	31.07%
RMSPE	12.18%	7.52%	38.26%

match the locations of merges as compared to the model that only captured "normal" lane-changing. Particularly, with the simple model, vehicles were unable to change lanes by accepting available gaps. Therefore, a large share of the merges occurred very late (81% occurred less than 100 meters from the end of the merging lane). This may result in the formation of queues on the ramp and an over-prediction of delays to both ramp and mainline vehicles. With the full model, the addition of the courtesy and forced merging mechanisms allowed drivers to merge more quickly and with greater ease, and thus only 47% of the merges occurred within 100 meters from the end of the ramp. This value is significantly closer to the observed 44%.

Comparisons of lane-specific flows from both versions of MITSIMLab are presented in Table 4.1. The results show that the full model was able to provide a better match to the actual flows. Thus, the improved realism of the model at the microscopic level was also translated into an improved fit to the aggregate (or macroscopic) traffic flow characteristics, which are most often the statistics of interest in a simulation application.

4.6   ACCOUNTING FOR HETEROGENEITY

All the lane-changing models discussed above incorporate decisions that drivers make over several choice dimensions (e.g., the choice of target lane, gap acceptance). Moreover, these decisions are repeated over time. Invariant characteristics of the drivers and their vehicles, such as aggressiveness, their level of driving skill and the vehicle's speed and acceleration capabilities, create correlations among the choices made by a given driver over time and choice dimensions. It is important to capture these correlations in the utility functions. However, the data available for model estimation does not comprehend information about these characteristics. Therefore, a model specification including individual-specific latent variables in the various utilities in order to capture these correlations was utilized. This individual-specific term appears in the utilities of all the various alternatives that a driver has in all his choices and in all time periods. The model assumes that, conditional on the value of this latent variable, the error terms of different utilities are independent. This specification is given by:

$$U_{int}^c = \beta_i^{c^T} X_{int}^c + \alpha_i^c v_n + \varepsilon_{int}^c \tag{4.1}$$

where U_{int}^c is the utility of alternative i of choice dimension c to individual n at time t. X_{int}^c is a vector of explanatory variable. β_i^c is a vector of parameters. v_n is an individual-specific latent variable assumed to follow some distribution in the population. α_i^c is the parameter of v_n. ε_{int}^c is a generic random term which is independently and identically distributed (i.i.d.) across alternatives, individuals and time. ε_{int}^c and v_n are independent of each other.

The resulting error structure (see Heckman, 1981; Walker, 2001 for a detailed discussion) is given by:

$$cov\left(U_{int}^c, U_{i'n't'}^{c'}\right) = \begin{cases} \left(\alpha_i^c\right)^2 + \left(\sigma_i^c\right)^2 & \text{if } n = n', \, c = c', \, i = i' \text{ and } t = t' \\ \alpha_i^c \alpha_{i'}^{c'} & \text{if } n = n', \, c \neq c' \text{ and/or } i \neq i' \text{ and/or } t \neq t' \\ 0 & \text{if } n \neq n' \end{cases} \quad (4.2)$$

where σ_i^c is the standard deviation of ε_{int}^c.

The impact of this formulation on the resulting behavior is demonstrated with the application in the lead and lag critical gaps of the merging model described in this model. Here, critical gaps depend on the remaining distance to the end of the merging lane. The individual specific random term was introduced in the coefficient of the remaining distance in both the lead and lag critical gaps (as well as in other parts of the model). This variable was interpreted as representing the range of drivers' behavior from timid to aggressive. Figure 4.11 demonstrates critical lead and lag gaps

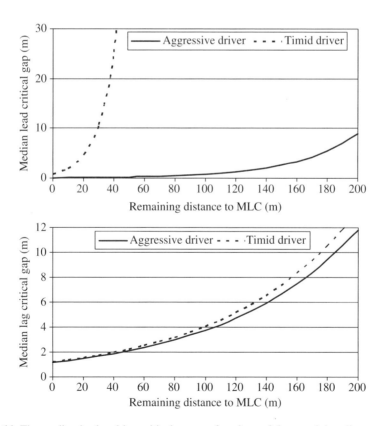

Fig. 4.11 The median lead and lag critical gaps as functions of the remaining distance to the end of the merging lane.

for timid and aggressive drivers. Everything else being equal, aggressive drivers had lower critical gaps as compared to their more timid counterparts. Furthermore, the difference in critical gaps increased when the merge became more urgent, as the vehicle approached the end of the merging lane. The use of the same individual-specific latent variable in both critical gaps also ensured that the behavior was consistent, and consequently, a driver who is aggressive in one dimension (e.g., has a small critical lag gap) would also be aggressive in other dimensions (e.g., would have a small critical lead gap as well).

4.7   CONCLUSIONS

Lane-changing is an important component of microscopic traffic simulation models, and has a significant impact on the results of analyses employing these tools. In recent years, the interest in the development of lane-changing models and their implementation in traffic simulators has increased dramatically. This chapter presents several enhancements to the basic lane-changing model that has been utilized, with some variations, in several simulators. These enhancements have been intended to form a more comprehensive modeling framework for the integration of various aspects of lane-changing behavior, such as MLC, DLC, and other lane changing mechanisms, including courtesy and forced gap acceptance. Estimation results and validation case studies demonstrated significant improvements in the ability of the enhanced lane-changing models to replicate observed behavior and traffic patterns as compared to the simple generic model. The extent of the improvement obtained with the enhancements presented herein leads us to believe that further advances in lane-changing models may give rise to additional improvements in their ability to replicate reality. In particular, two areas of improvement may be useful in that respect:

Integrating acceleration behavior in lane-changing models. Drivers' acceleration may be affected by their lane changing behavior. For example, drivers may accelerate or decelerate in order to position their vehicles such that they are able to accept available gaps. This type of behavior, if implemented in traffic simulators, may have a significant impact on simulated traffic flow characteristics. Some research in this direction, with promising results, has been conducted by Zhang et al. (1998) and Toledo (2002). However, further research to experiment with various model structures and specifications as well as to use more datasets from diverse locations and traffic conditions, is required in order to better understand the inter-dependencies among lane changing and acceleration behaviors.

Lane-changing behavior in arterial streets. All the results presented in this chapter and most of the research in this field in general are based on data collected from freeway sections. While the structures of these models are common enough to be applied on traffic in urban arterials, some factors affecting the lane-changing behavior in urban streets may not be present in freeway traffic. For example, the impact of buses and bus stops, paring activity, traffic signals and the queues that form behind them are important in urban streets but cannot be observed in data collected from freeway sections.

Furthermore, a significant proportion of lane changes in urban arterials may occur at intersections and not in the sections themselves. Research in this direction has been conducted by Wei et al. (2000). In an on-going effort sponsored by the NGSIM project, data collected in an arterial street in Los Angeles, California is used for this purpose.

4.8   ACKNOWLEDGMENTS

This chapter is based upon work supported by the National Science Foundation under Grant No. CMS-0085734 and by the Federal Highway Administration under contract number DTFH61-02-C-00036. Any opinions, findings and conclusions or recommendations expressed in this publication are those of the authors and do not necessarily reflect the views of the National Science Foundation or the Federal Highway Administration. Dr. Toledo is a Horev Fellow supported by the Taub and Shalom Foundations.

4.9   REFERENCES

Ahmed, K. I. (1999) *Modeling Drivers' Acceleration and Lane Changing Behavior*, PhD Dissertation, Department of Civil and Environmental Engineering, MIT.

Ahmed, K. I., Ben-Akiva, M., Koutsopoulos, H. N., and Mishalani, R. G. (1996) "Models of freeway lane changing and gap acceptance behavior", in Proceedings of the 13th International Symposium on the Theory of Traffic Flow and Transportation, pp. 501–515.

Choudhury, C., Ben-Akiva, M., and Toledo, T. (2007a) "Modeling lane-changing behavior in presence of exclusive lanes", Paper presented at the 11th World Conference on Transport Research, Berkley, USA.

Choudhury, C., Ben-Akiva, M., Toledo, T., Rao, A., and Lee, G. (2007b) "Modeling state-dependence in lane-changing behavior", Paper Presented at the 17th International Symposium on Transportation and Traffic Theory, London, UK.

Gipps, P. G. (1986) "A model for the structure of lane changing decisions", *Transportation Research Part B*, 20:403–414.

Halati, A., Lieu, H., and Walker, S. (1997) "CORSIM – corridor traffic simulation model", in Proceedings of the Traffic Congestion and Traffic Safety in the 21st Century Conference, pp. 570–576.

Heckman, J. J. (1981) "Statistical models for discrete panel data", in C. F. Manski and D. McFadden (Ed.), Structural Analysis of Discrete Data with Econometric Applications, pp. 114–178.

Hidas, P. (2002) "Modelling lane changing and merging in microscopic traffic simulation", *Transportation Research Part C*, 10:351–371.

Hidas, P., and Behbahanizadeh, K. (1999) "Microscopic simulation of lane changing under incident conditions", in Proceedings of the 14th International Symposium on the Theory of Traffic Flow and Transportation, abbreviated presentation sessions, pp. 53–69.

Lee, G. (2006) *Modeling Gap Acceptance in Freeway Merges*, MS Thesis, Department of Civil and Environmental Engineering, MIT.

Toledo, T. (2002) *Integrated Driving Behaviour Modelling*, PhD Dissertation, Department of Civil and Environmental Engineering, MIT.

Toledo, T. (2006) "Driving behaviour: models and challenges", *Transport Reviews*, 27(1):65–84.

Toledo, T., Ben-Akiva, M., and Koutsopoulos, H. N. (2003) "Modeling integrated lane-changing behavior", *Transportation Research Record*, 1857:30–38.

Toledo, T., and Koutsopoulos, H. N. (2004) "Statistical validation of traffic simulation models", *Transportation Research Record*, 1876:142–150.

Toledo, T., Koutsopoulos, H. N., Davol, A., Ben-Akiva, M., Burghout, W., Andreasson, I., Johansson, T., and Lundin, C. (2003) "Calibration and validation of microscopic traffic simulation tools: Stockholm Case Study", *Transportation Research Record*, 1831:65–75.

Walker, J. L. (2001) *Extended Discrete Choice Models: Integrated Framework, Flexible Error Structures and Latent Variables*, Ph. D. Dissertation, Department of Civil and Environmental Engineering, MIT.

Wei, H., Lee, J., Li, Q., and Li, C. J. (2000) "Observation-based lane-vehicle-assignment hierarchy for microscopic simulation on an urban street network", in Preprints of the 79th Transportation Research Board Annual Meeting, Washington D. C.

Yang, Q., and Koutsopoulos, H. N. (1996) "A microscopic traffic simulator for evaluation of dynamic traffic management systems", *Transportation Research Part C*, 4:113–129.

Zhang, Y., Owen, L. E., and Clark, J. E. (1998) "A multi-regime approach for microscopic traffic simulation", in Preprints of the 77th Transportation Research Board Annual Meeting, Washington D. C.

CHAPTER 5

TRAFFIC SIMULATION OF A RURAL 2 + 1 HIGHWAY IN HOKKAIDO

Kazunori Munehiro, Toshio Kamiizumi, Mamoru Sasaki, Toshiya Uzuka, Motoki Asano

In the strive toward service level improvement, road administrators have recently examined the development of "2 + 1 highways," in which some sections of an existing two-lane highway have been equipped with an auxiliary lane. Such improvements have been considered for national highways in rural Hokkaido, Japan. The present chapter investigates the lane-changing behavior on a 2 + 1 highway section using the SIM-R traffic flow micro-simulation program. The behavior of vehicles moving from the main lane to a left-hand auxiliary lane and back (in order for following vehicles to overtake them) has been observed since 2005 on a section of National Highway 38 in Shiranuka, Hokkaido. Based on these observation results, SIM-R was modified to include the lane-changing behavior with regard to giving way and passing.

5.1   INTRODUCTION

In recent years, road administrators have been examining service level improvement by developing "2 + 1" highways, in which various sections of an existing two-lane highways are equipped with an auxiliary lane. Such an improvement has been introduced on numerous sections of national highways in rural Hokkaido, Japan. In light of this, an appropriate traffic simulation program for rural 2 + 1 highways is required.

Lane-changing is a complex decision since it depends on a number of parameters. Gipps (1986) proposed a general model for lane-changing decisions for urban multiple-lane streets. The model places a high importance on the desired speed and the braking distance, and the factors governing the lane-changing decision process include the selection of lanes, the feasibility of changing lanes and the effect of the preceding vehicle. Toledo et al. (2003, 2005) proposed a basic lane-changing model based on the gap acceptance approach, in which distances between the subject vehicle and the vehicles immediately ahead and immediately behind the subject vehicle are incorporated. Tapani (2005) presented the RuTSim traffic microsimulation for rural road traffic. This model incorporates the overtaking decision-making process on a two-lane road, which is governed by four conditions: the vehicle's ability to overtake, the possibility of overtaking considering the surrounding traffic, and possible overtaking restrictions. Carlsson and Tapani (2006) simulated rural traffic in Sweden using the RuTSim traffic micro-simulation model in order to calibrate and verify the rural road simulation model. However, no studies on lane-changing behaviors have focused on the simulation of a rural 2 + 1 highway with the incorporation of a gap acceptance between the preceding vehicle and the subject vehicle in the "giving-way" lane, as well as that between the subject vehicle and the following vehicle in the "giving-way" lane.

An augmentation in the development of 2 + 1 highways has increased the need for a reliable model with respect to giving way on a rural 2 + 1 highway. Such a model could be calibrated by using observation data. To this end, the lane-changing model of SIM-R was improved by calibrating it with observation data. SIM-R is a traffic micro-simulation program that was developed by the former Civil Engineering Research Institute (CERI) in Japan in 1996. SIM-R is capable of simulating traffic flow for uninterrupted sections and for signalized sections, and is also able to forecast and evaluate summer and winter road traffic conditions. Thus, SIM-R is able to accurately reproduce and estimate traffic conditions.

The behaviors of vehicles moving from the main lane to a left-hand auxiliary lane (a "giving-way" lane) and back, in order to allow following vehicles to overtake them, were observed on a section of National Highway 38 in Shiranuka, Hokkaido, Japan in August 2005. The observations were carried out over a ten-day period using ten traffic counters and one video camera. The lane-changing model of the SIM-R traffic microsimulation program was calibrated based on the observation results, with the intention of reproducing and evaluating traffic flow on a 2 + 1 highway section.

The objectives of the study described in the present chapter were three-fold:

- to propose a lane-changing-behavior model for giving way and overtaking on a rural 2 + 1 highway, based on survey results on National Highway 38 in Shiranuka;

- to calibrate parameters of lane-changing behaviors for giving way by the SIM-R;

- to reproduce traffic by the SIM-R, using data from National Highway 38 in Shiranuka.

5.2   STUDY SITES

5.2.1   Outline of the field survey on National Highway 38 in Shiranuka

National Highway 38 is a major national highway that crosses Hokkaido in the east-west direction. Shiranuka Town is situated to the west of Kushiro, a large city in eastern Hokkaido (Fig. 5.1a). The road section that was surveyed is a left-hand auxiliary-lane

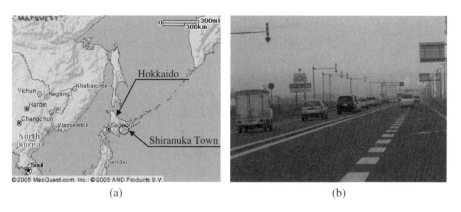

(a)   (b)

Fig. 5.1 (a) Hokkaido and Shiranuka town; (b) Photo of National Highway 38 in Shiranuka.

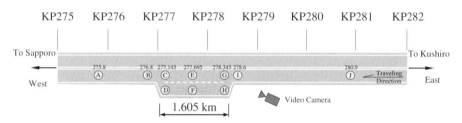

Fig. 5.2 The traffic counters installed at points A to J (National Highway 38 in Shiranuka).

section ($L = 1.605$ km) on National Highway 38 in Shiranuka (Fig. 5.1b). The section – a 2 + 1 lane, two-way road with a width of 15.0 m (each main lane: 3.50 m; the auxiliary lane: 3.0 m; the centerline: 1.0 m; each shoulder: 2.0 m) – runs through a flat area. Drivers in Japan drive to the left.

The survey was carried out for ten days from Friday, August 12 (7:00) to Sunday August 21 (7:00), 2005. The surveyed section was equipped with traffic counters (STC-2100P; Sumitomo 3M) to obtain data with regard to the speed, vehicle position and traffic volume of small vehicles (passenger cars) as well as large vehicles (trucks) in each lane. As shown in Figure 5.2, traffic counters were placed at ten locations from A to J. A video camera was installed at a fixed point in order to record vehicle changes from the main lane to the left-hand auxiliary lane and back.

5.2.2 Results of the field survey on National Highway 38 in Shiranuka

Speed distribution

A field survey (Munehiro et al., 2006) was conducted for all vehicle types, using traffic counters on the auxiliary-lane section (2 + 1 highway section) of National Highway 38 in Shiranuka during a time period of ten days, from August 12, 2005. Throughout this period, the surveyed road surface was mostly dry and the average daily traffic volume was 10,130 vehicles on weekdays and 11,815 vehicles on weekends. Large vehicles accounted for 13% of the daily traffic volume. Figure 5.3 displays the hourly

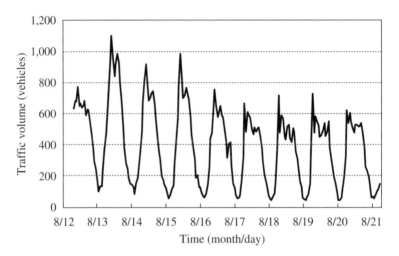

Fig. 5.3 The hourly traffic volume (Point I).

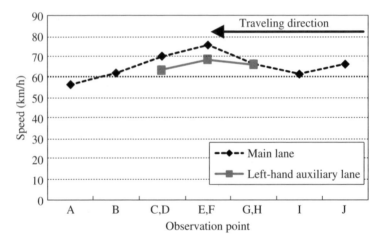

Fig. 5.4 The average speed at each observation point.

traffic volume during the survey period, representing approx. 600–800 veh/h during daytime (7:00–17:00) and more than 1,100 veh/h during the peak hours of August 13, 2005. The average speeds of the vehicles at each observation point and for each lane during the ten days are shown in Figure 5.4. Moreover, the speed distributions of small and large vehicles at Point J (main lane), Point E (main lane, midpoint of the 2 + 1 section) and Point F (left-hand auxiliary lane, midpoint of the 2 + 1 section) are shown in Figure 5.5. This figure suggests the following:

(1) The average speed of all vehicles was about 66 km/h at Point G (main lane, start of the 2 + 1 section) and Point H (left-hand auxiliary lane, start of the 2 + 1 section).

(2) For large vehicles, the average speed at Point E (the longitudinal midpoint of the main lane) was about 13 km/h higher than at Point F (the longitudinal midpoint

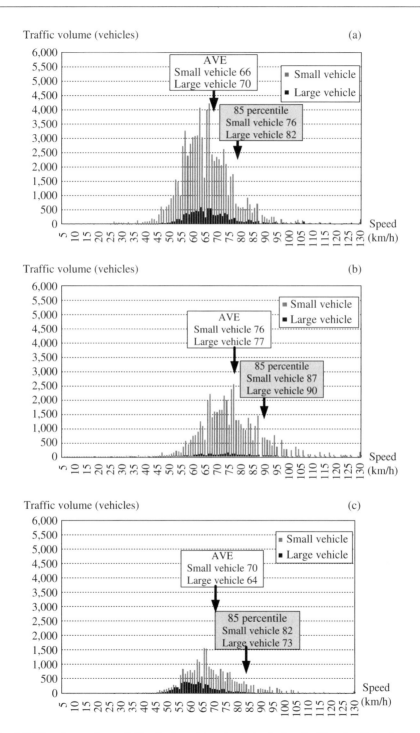

Fig. 5.5 The observed speed distribution: (a) point J (main lane); (b) point E (main lane); (c) point F (left-hand auxiliary lane).

Table 5.1 The use ratio of the main lane and the left-hand auxiliary lane.

	Total	Main lane	Change to left-hand auxiliary lane			
			Traveling independently	Leading platoon	Following in a platoon	Overtaking on the left lane
All	100% (1,004 vehicles)	51%	11%	22%	12%	4%
Small vehicle	100% (806 vehicles)	63%	9%	14%	10%	4%
Large vehicle	100% (198 vehicles)	6%	20%	52%	19%	3%

of the auxiliary lane). For small vehicles, the average speed at the former point was about 6 km/h higher than at the latter point.

(3) At Point E (the longitudinal midpoint of the main lane), the average observed speed of small vehicles was 76 km/h and the 85th-percentile speed was 87 km/h. This data indicates large variation in speeds.

Use of the left-hand auxiliary lane

Based on video images taken by a fixed camera, the traffic volume for the 2 + 1 section could be analyzed. This analysis (Sato et al., 2006) was conducted for 1,004 vehicles (806 small, 198 large) traveling between 10:00 and 12:00 on Thursday, August 18, 2005. Table 5.1 summarizes the use of the lanes. There were four categories of vehicles that changed to the left-hand auxiliary lane: those traveling independently, those leading a platoon, those following in a platoon and those overtaking on the left.

5.3   SIM-R TRAFFIC MICRO-SIMULATION PROGRAM

5.3.1   Simulation outline

Figure 5.6 presents the flow of the traffic simulation by SIM-R. On the basis of the abovementioned observations regarding vehicle behaviors on National Highway 38 in Shiranuka, the following three main simulation parameters were set:

(1) The traffic volume by vehicle type (small or large) in each lane was measured with the traffic counters. The input traffic observation data for simulation consisted of the traffic volume and the ratio of large vehicles.

(2) The speed distribution in each lane was also measured using the traffic counters. The data was used in two simulation cases: one for all traveling vehicles, and

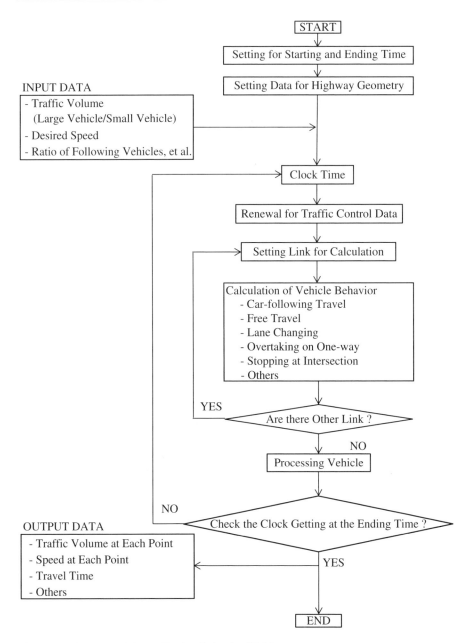

Fig. 5.6 A flowchart of the traffic simulation by SIM-R.

one for only free-traveling vehicles. The distribution of desired speeds was obtained from speed data of independent free-traveling vehicles collected at Points E and F, situated at the midpoint of the left-hand auxiliary-lane section. It was found that the desired speeds differed between the main lane and the left-hand auxiliary lane.

(3) The ratio of following vehicles was obtained through the following Inequality (5.1) (Japan Road Admin., 2004; Funamizu et al., 1992):

Vehicular gap (m) < Minimum braking distance of subject vehicle (m)   (5.1)

where

Vehicular gap (m) = time lag behind preceding vehicle (s) × V (m/s);

$$\text{Minimum braking distance (m)} = V \cdot t + \frac{V^2}{2gf}$$

V :   speed of subject vehicle (m/s);
t :   reaction time (lag in reaction by subject vehicle) (= 2.5 s);
f :   coefficient of sliding friction in the longitudinal direction;
g :   gravitational acceleration (=9.8 m/s²).

Vehicles traveling as expressed by the above inequality were defined as *following vehicles*, whereas all the others were defined as *free-traveling vehicles*. As road surface conditions generally vary between dry, wet, compacted snow and freezing, also the minimum braking distance varies.

5.3.2   Basic model for vehicle behaviors

Car-following model
Hermann's model, described by the equation below, is a typical car-following model on which the SIM-R program is based. The original model, however, does not consider vehicle length; instead, it assumes that the following vehicle uses the *front end* of the preceding vehicle to determine the minimum space headway, although a following vehicle actually uses the *rear end* of the preceding vehicle. Also, it was found that the length of the vehicle ahead of the subject vehicle should be taken into account in order to judge whether the subject vehicle was engaging in vehicle-following behaviors. Thus, the vehicular gap (space headway minus the length of preceding vehicle) was substituted as an input variable in the car-following model Eq. (5.2).

$$x''_{n+1}(T+t) = \alpha \frac{\left[x'_n(T) - x'_{n+1}(T) \right]}{\left[x_n(T) - x_{n+1}(T) \right]} \tag{5.2}$$

where

$x''_{n+1}(T+t)$:   acceleration of the following vehicle at time plus t seconds (m/s²);

$x'_n(T) - x'_{n+1}(T)$:   difference in speed between a preceding vehicle and a following vehicle (m/s);

$x_n(T) - x_{n+1}(T)$:   space headway (m);

α :   sensitivity coefficient (m/s);

t :   response time (=lag in reaction by following vehicle(s));

T :   time when simulation starts.

It was assumed that, for free traveling, each vehicle accelerated at its maximum acceleration until it reached its desired speed.

The SIM-R simulation program was modified to reproduce lane-changing behaviors on a $2+1$ highway section. Cases of lane changing included the following:

- changing from the main lane to the left-hand auxiliary lane;

- re-entering the main lane from the left-hand auxiliary lane.

A lane-changing behavior generally occurs when two conditions are fulfilled:

- the decision has been made to change lanes;

- lane-changing has been judged to be possible.

By using the modified SIM-R program, the traffic conditions on the $2+1$ section (left-hand auxiliary lane) of National Highway 38 in Shiranuka could be reproduced for comparison with the results of the abovementioned field survey.

5.4 MODEL OF LANE-CHANGING BEHAVIOR ON A RURAL $2+1$ HIGHWAY

Figure 5.7 depicts the flowchart process of changing lanes on a road section with a left-hand auxiliary lane.

5.4.1 Changing from the main lane to the left-hand auxiliary lane

Parameters governing the decision whether to change from a main lane to a left-hand auxiliary lane in the SIM-R traffic flow simulation program were modified as follows.

The decision to change lanes is finalized when the below conditions are fulfilled:

(1) The distance from the following vehicle $\leq$ the minimum braking distance of the following vehicle (when the following vehicle is part of a platoon). It is assumed that the decision to change to the left-hand auxiliary lane is made when the vehicle behind the subject vehicle changes from being free traveling to following, i.e., when the vehicular gap between the subject vehicle and the following free-traveling vehicle becomes less than the minimum braking distance of the following vehicle.

(2) The desired speed of the following vehicle – the desired speed of subject vehicle $\geq B_1$ (km/h). It is assumed that the subject vehicle is likely to decide to move into the left-hand auxiliary lane when the desired speed of the following vehicle exceeds that of the subject vehicle. It is also assumed that a vehicle is unlikely to move into the left-hand auxiliary lane when its desired speed exceeds that of the following vehicle.

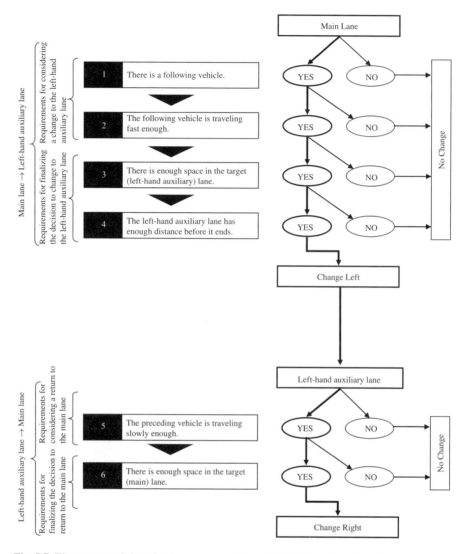

Fig. 5.7 The process of changing lanes on a section with a left-hand auxiliary lane.

Therefore, the difference in desired speed between the subject vehicle and the following vehicle (speed difference) is used for the simulation. The speed difference B_1 in the above inequality is a parameter used in reproducing the traffic flow.

The conditions for finalizing the decision to change to the left-hand auxiliary lane were determined as follows.

Vehicular gap

It is assumed that a vehicle finalizes its decision to change into the left-hand auxiliary lane if the expected vehicular gaps between it and the preceding and following vehicles in the left-hand auxiliary lane (i.e., D_1 and D_2 in Fig. 5.8) are equal to or

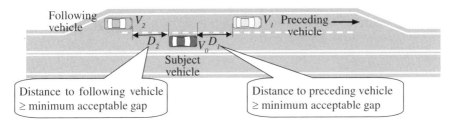

Fig. 5.8 The calculation of acceptable gaps when changing to the left-hand auxiliary lane.

exceed the minimum acceptable gaps. D_1 and D_2 are given by the right sides of the inequalities in Eq. (5.3).

$$D_1 > \frac{V_0^2}{2gf} + (V_0 - V_1)t + L_0 \quad \text{and} \quad D_2 > \frac{V_2^2}{2gf} + (V_2 - V_0)t + L_0 \qquad (5.3)$$

where

D_1 : distance between the preceding vehicle and the subject vehicle (m);
D_2 : distance between the subject vehicle and the following vehicle (m);
V_0 : speed of subject vehicle (m/s);
V_1 : speed of preceding vehicle (m/s);
V_2 : speed of following vehicle (m/s);
f : coefficient of sliding friction in the longitudinal direction;
L_0 : min. acceptable gap (≥ 1.5 m);
t : response time (lag in reaction by following vehicle)(=2.5 s);
g : gravitational acceleration (=9.8 m/s^2).

Passing
A consideration of whether a following vehicle is able to overtake the subject vehicle by the end of the left-hand auxiliary lane is added to the conditions of the lane-changing behavior. This is done to reproduce the behavior of vehicles that had decided to give way just before the end of left-hand auxiliary lane, but ended up not giving way. Such behavior is determined according to the following inequality: For a subject vehicle having reached the end of the left-hand auxiliary lane, the gap between that vehicle and the vehicle that overtook it must be greater than the minimum acceptable gap.

5.4.2   Re-entering the main lane from the left-hand auxiliary lane

Parameters governing the decision whether to consider re-entry into the main lane from the left-hand auxiliary lane in the SIM-R traffic flow simulation program were modified as follows:

(1) After changing to the left-hand auxiliary lane, the subject vehicle travels in this lane until it ends.

(2) The subject vehicle re-enters the main lane if there is a slower preceding vehicle in the left-hand auxiliary lane.

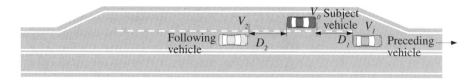

Fig. 5.9 The calculation of acceptable gaps when changing to the main lane.

As a basic rule, a vehicle that has moved into the left-hand auxiliary lane stays there until it the lane ends. If a slower vehicle is traveling ahead in the left-hand auxiliary lane, the decision of the subject vehicle of whether to re-enter the main lane follows the overtaking algorithm.

Conditions for finalizing re-entry into the main lane were set as follows:

(1) Distances to preceding and from following vehicles in the main lane must be greater than or equal to the minimum acceptable gap.

(2) In a situation where the subject vehicle moves to the main lane and encounters a platoon, the following vehicle must reduce its speed to secure the minimum acceptable gap.

The acceptable gaps to the preceding vehicle and from the following vehicle in the main lane required for re-entering the main lane are shown in D_1 and D_2 in Figure 5.9. In a situation where the subject vehicle joins a platoon in the main lane, it is assumed that the following vehicle reduces its speed to maintain the minimum acceptable gap. So as to secure a large enough vehicular gap from a following vehicle, the subject vehicle activates its turn indicator in order for the following vehicle to reduce its speed. When the minimum acceptable gap between the subject vehicle and the following vehicle is obtained, a maneuver to re-enter the main lane is executed.

5.5   SIMULATION RUN

5.5.1   Simulation requirements

Traffic data obtained between 10:00 and 11:00 on the morning of August 13, 2005 was employed in the simulation analysis. The traffic volume was calculated to be 1,100 veh/h, with large vehicles accounting for 7.7% of all vehicles, as estimated from the field observation data. The distribution of desired speeds was obtained from the speed data of independent free-traveling vehicles collected at Points E and F, which were situated at the midpoint of the left-hand auxiliary-lane section (Fig. 5.10). The speed distributions of all vehicles and independent free-traveling vehicles are presented in Figure 5.11.

The coefficient of the sliding friction, used to calculate the braking distance, was set at 0.60, which corresponds to a dry road surface. Other parameters are listed in Table 5.2, and the requirements for running the simulation can be found in Table 5.3. The values of maximum acceleration, maximum deceleration and vehicle length were

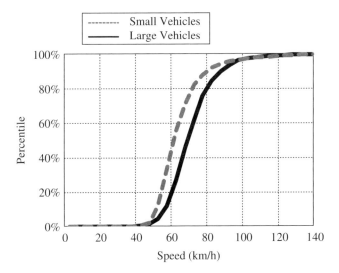

Fig. 5.10 The percentile as a function of the desired speed, taken from Points E and F on National Highway 38 in Shiranuka.

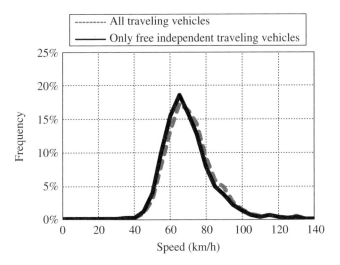

Fig. 5.11 The speed distribution taken from points E and F on National Highway 38 in Shiranuka.

set using Japan Road Administration (2004). To improve the accuracy of the simulation, a computing step time of 0.5 seconds was selected.

5.5.2   Simulation results

Based on the above conditions, simulations were run to reproduce the traffic conditions of the 2 + 1 highway section on National Highway 38 in Shiranuka. The survey was carried out for a section with a left-hand auxiliary lane and a dry road surface. Figures 5.12 and 5.13 show the relations between the simulated hourly traffic volumes

Table 5.2 Other simulation parameters.

Item	Set value
Max. accel.	
small vehicle	6 km/h/s
large vehicle	4 km/h/s
Max. decel. (Small and large vehicles)	17.6 km/h/s
Vehicle length	
small vehicle	4.7 m
large vehicle	12.0 m
Min. acceptable gap (Small and large vehicles)	1.5 m
Requirement for free traveling or following	Vehicular gap < Min. braking distance
Sensitivity coefficient	
accel.	8.2 m/s
decel.	17.0 m/s
Computing cycle time	0.5 s
B_1 (Desired speed of following vehicle – Desired speed of subject vehicle)	15 km/h

Table 5.3 The requirements for running the simulation.

Item	Required value	
Simulation repetitions[1]	5 times (to calculate an average)	
Initialization time (pre-simulation time)[2]	600 s	(10 min.)
Simulation duration	3,600 s	(60 min.)

[1]Simulation repetitions: Since each simulation run produces a different traffic pattern that affects the survey result, the simulation result was taken as the average of five runs.
[2]Initialization time: Time required to initialize the system in order to re-start the simulation.

and those observed as well as the relations between the percentile of simulated speed and that of the observed speed for points C to I. The function given by linear regression analysis of traffic volume vs. simulated traffic volume had a gradient within 1 ± 0.1 for observation points C to I. Moreover, the coefficients of determination exceeded 0.9.

The results indicate that the simulation accurately reproduced the lane-changing behavior of the studied section. The reproducibility was slightly lower at observation point C, situated close to the end of the auxiliary lane.

(1) The speed distribution was accurately reproduced at observation points D to I. However, it was lower for Point C.

(2) The simulation parameter B_1 (i.e., the desired speed of the following vehicle – the desired speed of the subject vehicle) was 15 km/h. At points E to I, the reproducibility for the traffic volume and speed distribution was high.

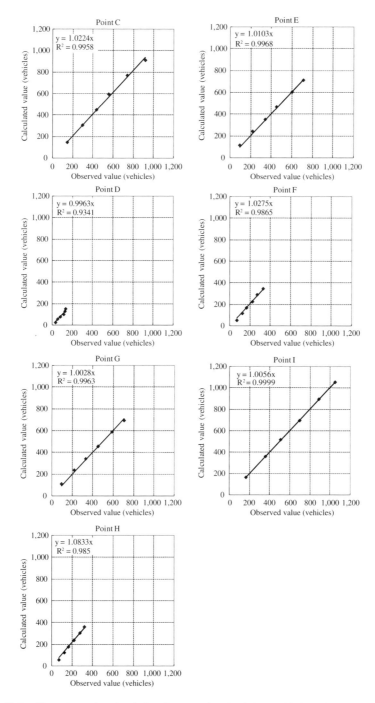

Fig. 5.12 Traffic volume by simulation from National Highway 38 in Shiranuka under dry conditions and with 1,100 vehicles/hour.

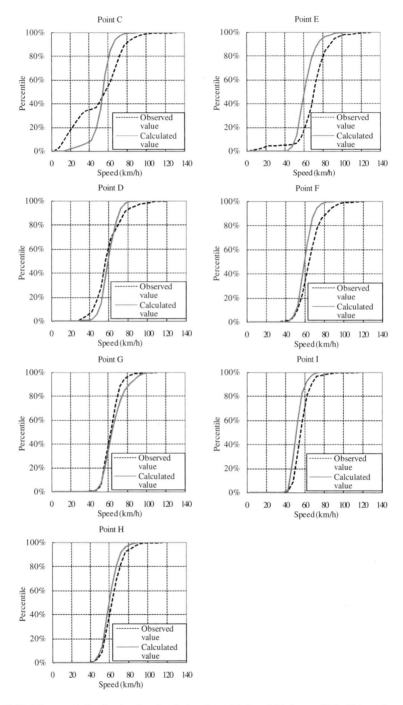

Fig. 5.13 The speed distribution by simulation from National Highway 38 in Shiranuka under dry conditions and with 1,100 vehicles/hour.

5.6   DISCUSSION AND CONCLUSIONS

5.6.1   Field survey results for National Highway 38 in Shiranuka

A survey on vehicle behaviors on an auxiliary-lane section (2 + 1 highway section) of National Highway 38 in Shiranuka was carried out during a time period of ten days in August 2005. A traffic counter was installed at ten observation points. The highest observed average speed for all vehicles on both the main lane and the auxiliary lane was found at the midpoint of the auxiliary-lane section. The average speed for all vehicles in the main lane was more than 10 km/h greater than that in the left-hand auxiliary lane. Based on two hours of video images taken with a fixed video camera, an analysis was performed on the use of the left-hand auxiliary lane. Vehicles that used the left-hand auxiliary lane accounted for 49% of all vehicles, whereas the remaining 51% used the main lane. As for large vehicles, 94% of them traveled in the left-hand auxiliary lane.

5.6.2   Lane-changing behavior model of rural 2 + 1 highways

The behavior of changing to the left-hand auxiliary lane and re-entering the main lane was simulated using the vehicular gaps between a preceding vehicle and the subject vehicle, as well as between the subject vehicle and a following vehicle traveling in the left-hand auxiliary lane. Based on the results, the program was modified to simulate lane-changing behaviors.

The conditions for lane-changing were divided into two categories: the conditions for considering a change, and the conditions for finalizing the decision to change. The decision depended on the gaps between the subject vehicle and the preceding and following vehicles. It was assumed that a subject vehicle was likely to decide to move into the left-hand auxiliary lane when the desired speed of the following vehicle exceeded that of the subject vehicle. Parameter B_1 provided the difference between the desired speed of the following vehicle and that of the subject vehicle. Subsequently, the lane-changing model and parameter B_1 were incorporated into the SIM-R traffic micro-simulation program.

5.6.3   Calibration of parameters of the SIM-R traffic micro-simulation for the lane-changing behavior

Three main parameters were used in the SIM-R traffic micro-simulation program: the traffic volume in each lane for two vehicle types, the distribution of the desired speed for each lane, and the ratio of following vehicles. Parameter B_1 was defined as the difference in desired speeds between the subject vehicle and the preceding and following vehicles, and it was set to 15 km/h. By using the SIM-R program, the traffic was simulated for the auxiliary-lane section (2 + 1 highway section) of National Highway 38 in Shiranuka, under conditions of a dry road surface.

The correlation between the simulated and observed spot traffic volume was high. The function approximated by liner regression analysis of the observed traffic volume vs. the simulated traffic volume displayed differences within 1 ± 0.1, and the coefficient of determination exceeded 0.9 at each observation point.

With the exception of at observation point C, the speed percentiles from the observed data agreed well with that of the simulated data. At Point C, on the other hand, the observed vehicle speeds tended to be lower than their simulated vehicle counterparts. This was attributed to Point C's proximity to the merging point between the auxiliary lane and the main lane.

Future work includes plans to further modify the model based on the results of additional field surveys conducted on wet roads. Furthermore, for use in cold and snowy conditions, also winter field surveys will be conducted in order to modify the model such that it will reproduce lane-changing behaviors for roads covered with ice and compacted snow. In this way, the model will be expanded to achieve year-round applicability thus enabling the reproduction of year-round traffic flows. Ultimately, this should lead to enhanced designs for rural 2 + 1 highways in cold, snowy regions.

5.7   ACKNOWLEDGMENTS

This report partially incorporates results of the 2005 Practical ITS Study (A-3) of the Japan Society of Civil Engineers. Special thanks are expressed to Dr. Tetsuo Shimizu of Tokyo University, Dr. Kiyoshi Takahashi of Kitami Institute of Technology, the Hokkaido Development Bureau, the MLIT and the all the members of the Hokkaido study team.

5.8   REFERENCES

Carlsson, A., and Tapani, A. (2006) "Rural highway design analysis through traffic micro-simulation", *Proceedings of the 5th International Symposium on Highway Capacity and Quality of Service*, 2:249–258.

Funamizu, M. et al., (1992) "Development of traffic simulation model for ordinary roads", *The 12th Proceedings of the Japan Society of Traffic Engineers*.

Gipps, P. G. (1986) "A model for the structure of lane-changing decisions", *Transportation Research Board*, 20B(5):403–414.

Japan Road Administration (2004) Explanation and Enforcement of Road Structure Act. Japan.

Munehiro, K. et al., (2006) "Study on highway geometric design considering regional and climatic features of Hokkaido", *33rd Proceedings of Infrastructure Planning and Management*.

Sato, K. et al. (2006) "An analysis of vehicle's maneuvers on give-way lane section", *34th Proceedings of Infrastructure Planning*.

Tapani, A. (2005) "Versatile model for simulation of rural road traffic", *Transportation Research Record*, 1934:169–178.

Toledo, T., Choudhury, C., and Ben-Akiva, M. (2005) "Lane-changing model with explicit target lane choice", *Transportation Research Record*, 1934:157–165.

Toledo, T., Koutsopoulos, H., and Ben-Akiva, M. (2003) "Modeling integrated lane-changing behavior", *Proceeding of 82nd TRB Annual Meeting*.

TRB. (2005) *Highway Capacity Manual 2000*.

CHAPTER 6

MECHANICAL RESTRICTION VS. HUMAN OVERREACTION: MODELING OF SYNCHRONIZED TWO-LANE TRAFFIC

Andreas Pottmeier, Christian Thiemann, Andreas Schadschneider, Michael Schreckenberg

This contribution presents a realistic cellular automaton model for multi-lane traffic, based on the model proposed by Lee et al. (2004). In contrast to current approaches, a limited deceleration capability, i.e., a mechanical restriction, has been assigned to the vehicles. Moreover, the velocity of the vehicles was determined on the basis of the local neighborhood, and the drivers were thus divided into optimistic or pessimistic drivers. The former were prone to underestimating their safety distance if their neighborhood admitted it, whereas the latter always kept a safe distance, thus overreacting. This resulted in a convincing reproduction of the microscopic and macroscopic features of synchronized traffic. The anticipation of the leader's velocity was thereby essential for the reproduction of synchronized traffic. Nevertheless, accidents occurred in the stationary state, and thus, the model approach required modification so as to be capable of simulating multi-lane traffic. The adapted model was enhanced by a realistic lane-change algorithm, and a multi-lane model, reproducing the empirical data even better than the single-lane approach, was formulated. In open systems with bottlenecks, such as an on-ramp or a speed-limit, the empirically observed complex structures of the synchronized traffic could be reproduced in great detail.

6.1 INTRODUCTION

One objective of current traffic research is the realistic description of vehicular traffic by means of modeling. The term "realistic" is used in the sense that macroscopic and microscopic empirical data are reproduced with a high degree of realism (Knospe et al., 2004). Since no present traffic model is capable of reproducing all empirical characteristics, the development of such a model is of great interest.

In recent times, various approaches for the modeling of complex traffic features, especially the emergence of synchronized traffic (Kerner and Klenov, 2002; Kerner et al., 2002) and the pinch effect, have been proposed. In 2004, Lee et al. (2004, 2005) introduced an advanced cellular automaton model for single-lane traffic which reproduced the various forms of synchronized traffic, the pinch effect (Kerner, 1998) as well as short time-headways in free flow (Kerner and Klenov, 2002). In the model, realistic flow properties are consequences of moderate driving, and it takes into account finite acceleration and deceleration properties in addition to the attitude of the drivers.

In order to simulate the various behaviors of drivers, one distinguishes between optimistic and pessimistic driving. These two cases correspond to scenarios in which the vehicles either drive unhindered and drivers accept "unsafe" gaps, or interactions between the cars are strong and braking is likely.

Due to the finite deceleration capabilities, the model is not intrinsically accident-free. In the presence of limited deceleration capabilities, crashes have to be avoided or at least reduced by an appropriate choice of dynamics in order for the overall dynamic not to be altered. In the original version of the model (Lee et al., 2004), accidents could occur if the initial state was not carefully chosen. Therefore, an extension of the model increasing its robustness against accidents was required.

Thus, in parallel to the development of the rule-set for a secure lane change, the original one-lane model was adapted. Here the calculation scheme of the parameter γ, that describes the attitude of the driver, had to be changed. The characteristics of the accidents observed in simulations have shown that a delayed change from the optimistic to the pessimistic state is the source of the accidents in the original model.

6.2   MODEL DEFINITION OF THE SINGLE LANE MODEL

As a first step, the rule-set of the single-lane model is recapitulated. The calculation of the vehicles is accomplished by an inequality of which the purpose is to compute a maximum velocity c_n^{t+1} so that the follower can drive as fast as possible but still be able to stop behind his leader:

$$x_n^t + \Delta + \sum_{i=0}^{\tau_f(c_n^{t+1})} (c_n^{t+1} - Di) \le x_{n+1}^t + \sum_{i=1}^{\tau_l(v_{n+1}^t)} (v_{n+1}^t - Di) \qquad (6.1)$$

Here, x_n^t and v_n^t are the position and the velocity of vehicle n, respectively. Δ represents the minimum gap between the vehicles that is at least a distance l_{n+1} from the leading vehicle. D denotes the deceleration capability of the vehicles, $\tau_f(c_n^{t+1})$ and $\tau_l(v_n^{t+1})$ describe the number of time-steps in which the velocity is calculated and the safety is ensured for the follower and the leader (see Eq. (6.3)).

Each summation in Eq. (6.1) denotes a deceleration cascade with a maximum braking capability D. As long as both $\tau_f(v)$ and $\tau_l(v)$ are set to v/D and $\Delta = L$, the deceleration would end in a bumper-to-bumper configuration. But this is weakened by the introduced human factor that evaluates the neighborhood, i.e., the next two vehicles in front (see below), and acts depending on the results. Note that c_n^{t+1} is determined by the upper limit of Eq. (6.1). Nevertheless, the retarded reaction, i.e., the vehicle responding to its leader one time step later, of vehicle n is modeled by the different lower summation limits. If $\tau_l < 1$, the sum to the right is equal to zero while the sum on the lefthand side always has at least one term, i.e., c_n.

In order to model the various behaviors of the drivers, not only the usual stochastic variation of the driving is applied. The model introduces an attitude of the driver and distinguishes between optimistic and pessimistic driving. The vehicles drive with respect to their neighborhood, i.e., the driver considers the two vehicles ahead of himself. Furthermore, the driver evaluates the local situation, represented by the Boolean variable, and decides whether he should drive optimistically $\gamma = 0$ or pessimistically $\gamma = 1$.

The former controls the driving behavior in free flow: the vehicles drive unhindered and drivers accept "unsafe" gaps, i.e., gaps smaller than those allowing them to react to an emergency braking of the leading vehicle. Short time headways of less than one second are possible (Knospe et al., 2002). The latter parameter governs the driving at high densities. The interaction between cars is high and braking is likely, even overreacted braking maneuvers. The vehicles drive pessimisticly and remain aloof. This leads to the following specification of γ.

$$\gamma_n^t = \begin{cases} 0 & \text{for } v_n^t \le v_{n+1}^t \le v_{n+2}^t \text{ or } v_{n+2}^t \ge v_{\text{fast}} \\ 1 & \text{otherwise} \end{cases} \tag{6.2}$$

Here, v_{fast} is a high velocity that is slightly below the maximum velocity: $v_{\text{fast}} = v_{\text{max}} - 1$. This definition gives two reasons for a driver to act optimistically: he drives in a platoon of vehicles that either have the same speed or are driving faster than himself, i.e., $v_n^t \ge v_{n+1}^t \ge v_{n+2}^t$, or the vehicle that is two cars ahead is speeding away with high velocity, $v_n^t \ge v_{n+1}^t \ge v_{\text{fast}}$. Both reasons describe a local situation in which the driver is not likely to be hindered by its leaders such that he has to anticipate a hard braking action.

The calculation of the upper limits $\tau_l(v)$ and $\tau_f(v)$ of the summation in Eq. (6.1) as well as the minimum gap between the cars are as follows:

$$\Delta = L + \gamma_n^t \max\{0, \min\{g_{\text{add}}, v_n^t - g_{\text{add}}\}\}$$
$$\tau_f(v) = \gamma_n^t v/D + (1 - \gamma_n^t) \max\{0, \min\{v/D, t_{\text{safe}}\} - 1\}$$
$$\tau_l(v) = \gamma_n^t v/D + (1 - \gamma_n^t) \min\{v/D, t_{\text{safe}}\} \tag{6.3}$$

Here, D denotes the deceleration capability of the vehicles and t_{safe} the maximum number of time steps a vehicle observes its own safety when driving optimistically. The driver monitors the safety of his situation only towards this limited "time horizon". g_{add} is an additional safety gap (to avoid accidents) for the optimistic drivers ($\gamma_n^t = 0$) that drive faster than $2 \cdot g_{\text{add}}$.

The remaining update is as follows:

$$p = \max\{p_d, p_0 - v_n^t(p_0 - p_d)/v_{\text{slow}}\}$$
$$\tilde{c}_n^{t+1} = \max\{c_n^{t+1} \mid c_n^{t+1} \text{ satisfies Eqs. (6.1)-(6.3)}\}$$

$$\tilde{v}_n^{t+1} = \max\{0, v_n^t + a, \max\{0, v_n^t - D, \tilde{c}_n^{t+1}\}\}$$
$$v_n^{t+1} = \max\{0, v_n^t - D, \tilde{v}_n^{t+1} - \eta\}, \text{ with } \eta = 1 \text{ if rand}() < p \text{ or } 0 \text{ else}$$
$$x_n^{t+1} = x_n^t + v_n^{t+1} \tag{6.4}$$

The calculation of the stochastic parameter p is a generalization of the slow to start rule (see Barlović et al., 1998, 2002; Knospe et al., 2000) that interpolates linearly between p_d and p_0 for vehicles driving slower than v_{slow}, whereas slow vehicles are more likely to dawdle. This is important when reflecting the compact jam patterns that are often found in congested traffic flows. It has previously been shown Barlović et al. (2001), that this ingredient produces stripped jam patterns in simple cellular automaton models. Note that the limited deceleration capability also limits the dawdling of a vehicle (Eq. 6.4), i.e., the vehicle cannot dawdle when it has reduced its velocity maximally.

The second step of the update algorithm calculates the maximum safety velocity $\tilde{c}_n^{t+1}$. Here, the mechanical restrictions are applied and the traffic is regulated. After the dawdling step, the vehicle is moved. It should be noted that the approach is not intrinsically accident free.

6.3   STABILITY OF THE MODEL

An analysis of the model has demonstrated that it suffers from accidents (see also Lee et al., 2004), and an in-depth analysis of the mechanisms that are responsible for the accidents is thus required. This is particularly important for the development of a sophisticated lane change algorithm.

It has been mentioned (Lee et al., 2004), that the model may suffer from accidents if vehicles are carelessly inserted at an on-ramp, however, analyses show that such careless insertions of vehicles are not the only source for accidents. A crash may also occur in a periodic system that is initialized as claimed in Lee et al. (2004). Thus, for two-lane traffic or for the introduction of on-ramps, the model has to be adapted. The analysis of the accidents – no accidents were detected in the free flow region whereas, at higher densities (e.g. $\sigma = 60$ veh/km), a maximum number of accidents were found, in which a vehicle had an accident every 10^8 time-steps corresponding to 12.000 days of real time – showed that critical situations emerged at the transition between optimistic and pessimistic driving as the vehicle reacted with an offset of one time step.

Simulations have demonstrated that two different scenarios lead to a dangerous configuration. The first is a case where all three vehicles, i.e., vehicles n, $n + 1$ and $n + 2$, that are involved in the calculation of γ_n drive optimistically with the same velocity. If this platoon approaches vehicles driving substantially slower, it may occur that vehicle n reacts too late to the deceleration of the two leading vehicles. The second case is when the velocity difference between v_{n+1} and v_{n+2} is too high. This will not happen in a system with a single lane but is valid when vehicles are inserted because of an on-ramp or if the vehicles change lanes in a two-lane system.

Due to the nature of the collisions observed in the original model, the definition of γ_n was changed. Nevertheless, it is still possible to define a model without any accidents, and this requires the addition of a stronger interaction between each vehicle n and $n + 2$ as well as the prevention of the formation of the platoons mentioned above.

The former can be achieved by introducing a brake-light b_n that denotes whether the vehicle has reduced its velocity because of its surrounding, but not because of dawdling (i.e., the randomization). Moreover, this factor sets the attitude to pessimistic if one of the leading vehicles $n + 1$ or $n + 2$ was found to brake in the previous time-step. The latter could be done by changing the first inequality in the calculation of γ to $\gamma_n \leq \gamma_{n+1} < \gamma_{n+2}$ (Pottmeier et al., 2007).

$$b_n^t = \begin{cases} 1 & \text{for } \tilde{v}_n^{t+1} < v_n^t \\ 0 & \text{otherwise} \end{cases}$$

Since these changes would lead to slightly different dynamics of the model in the higher density region where the synchronized phase is unrealistically stabilized, another approach is followed. The update algorithm, and especially the calculation of γ, is altered so that no additional accidents are provoked because of vehicles changing lanes. This leads to a different computation of γ:

$$\gamma_n^t = \begin{cases} 0 & v_n^t \leq v_{n+1}^t \leq v_{n+2}^t \text{ or} \\ & (v_{n+2}^t \geq v_{\text{fast}} \text{ and } v_n^t - v_{n+1}^t < D \text{ and } gap_n < 8) \\ 1 & \text{otherwise.} \end{cases} \qquad (6.5)$$

The remaining update is unchanged.

The differences in the definition of the model dynamics hardly influence the macroscopic and microscopic results. The dynamics of the model are still capable of reproducing the traffic patterns observed empirically as well as the *short time-headways* in the free flow phase and *synchronized traffic*.

The following model parameters, which are motivated by empirical facts and already utilized in (Lee et al., 2004), are employed in the simulations: $D = 2$ cells/time-step[2], $v_{\text{fast}} = 19$ cells/time-step, $v_{\text{max}} = 20$ cells/time-step, $p_0 = 0.32$, $p_d = 0.11$, $t_{\text{safe}} = 3$ s, $g_{\text{add}} = 4$ cells, and $v_{\text{slow}} = 5$ cells/time-step. The length of a cell is set to $\Delta x = 1.5$ m and one time-step to $\Delta t = 1$ s.

6.4   TWO-LANE TRAFFIC

In addition to the modifications mentioned above, simulating realistic scenarios requires an extension to a two-lane model. Here, it is especially important to keep in mind the missing hardcore repulsion. The lane changes have to take into account safety in the sense of not interfering with cars in the other lane, but also by avoiding accidents in the process of changing lanes.

As the velocity calculation, and therefore also the determination of the security of the vehicles, does not only depend on the gap and the velocity of the involved vehicles but additionally requires the calculation of the inequality (see Eq. (6.1)), the lane change process has to be performed virtually. This means that the changing vehicle is virtually set to its destination lane. At this point, the velocity can be calculated and thus the security of the involved vehicles is determined in an intermediate step. The lane-changing rules are described in the next section following this approach and stating it in greater detail. Symmetric lane changes are considered as they qualitatively show the important features of the two-lane dynamics.

6.5   TWO-LANE MODEL

In the two-lane model, each time-step is separated into two sub-steps (Chowdhury et al., 2000). In the first step, it is decided whether each vehicle will change lanes in the current time-step. In the second step, the normal one-lane model update is applied to each of the two lanes.

The basic idea behind the two-lane model is to employ the condition for $\tilde{c}_n^{t+1}$ in order to determine how safe a possible lane change would be. In each time-step, it is verified whether the vehicle would be able to drive with the safe velocity if it changes lanes. It is also determined whether the follower on the destination lane would be able to drive safely. The parameter $\beta \in \{0,\ldots,D\}$ controls to what extent the follower may be constrained. For $\beta = 0$, only smooth lane changes are allowed where the follower is not forced to brake at all, while at $\beta = D$ even decelerations with a maximal braking capacity are acceptable. These conditions constitute the *security criterion* and determine whether a vehicle is *able* to change lanes without obstructing vehicles on the destination lane or even provoking a dangerous situation.

A vehicle will *want* to change lanes only if the *mobility criterion* is satisfied. For the symmetric two-lane model, this means that the vehicle can drive faster on the destination lane than on its current lane, which is determined by calculating $\tilde{v}_n^{t+1}$ on the current lane and then virtually changing the vehicle to the other lane, calculating $\tilde{v}_n^{t+1}$ again and comparing the two results.

According to Sparmann (1978), a lane change takes $t_{lc} = 3$ seconds, and the security criterion must therefore hold for at least three time-steps until a positive mobility criterion can trigger the vehicle to actually change lanes. For each vehicle, a new variable ϑ_n^t is introduced, which is initially $\vartheta_n^t = 0$ and acts as a counter for the time-steps in which the security and the mobility criteria become valid.

To formally describe the update rules of the model, $l_n^t \in \{0,1\}$ denotes the lane used by vehicle n at time t. $F_l(n)$ designates the follower and $L_l(n)$ the leader of the vehicle n on lane l. Some values are calculated for the virtual lane change of a vehicle, in which case, a second subscript is used to specify the lane. Thus $\tilde{v}_{n,l}^{t+1}$ means "$\tilde{v}_n^{t+1}$", if the vehicle would be on lane l.

The update rules of the two-lane model are, in turn, separated into two substeps. First, it is verified whether the vehicle would benefit from a lane-change, i.e., whether the mobility criterion is satisfied:

$$\tilde{v}_{n,1-l_n^t}^{t+1} > \tilde{v}_{n,l_n^t}^{t+1} \tag{6.6}$$

This means that a vehicle tends to change lanes if the velocity it can reach on the destination lane is higher than that on its original lane. To ensure a safe lane-change, the security criterion also needs to be valid:

$$x_n^t - x_{\mathcal{F}_{1-l}(n)}^t > L + g_{safe}$$

$$x_{\mathcal{L}_{1-l}(n)}^t - x_n^t > L + g_{safe}$$

$$v_n^t - D \leq \tilde{v}_{n,1-l_n^t}^{t+1}$$

$$v_{\mathcal{F}_{1-l_n^t}(n)}^t - \beta \leq \tilde{v}_{\mathcal{F}_{1-l_n^t}(n)}^{t+1} \qquad (6.7)$$

Here, g_{safe} denotes an optional security gap that becomes important at higher densities where the gap between the vehicles approaches zero.

If, as a result, a safe lane-change is possible (and wanted) ϑ_n is increased, $\vartheta_n^{t+1} = \vartheta_n^t + 1$, and a verification is carried out to see whether the conditions Eqs. (6.6) and (6.7) were able to hold t_{lc} times. If this is the case, the vehicles changes lanes:

$$t_n^{t+1} = 1 - t_n^t \text{ if } \vartheta_n^{t+1} = t_{lc} \text{ or } l_n^t \text{ otherwise,}$$

$$\vartheta_n^{t+1} = 0 \text{ if } l_n^{t+1} \neq l_n^t. \qquad (6.8)$$

Note that t_{lc} is important for the number of lane changes N_{lc}. For a detailed discussion of the influence of t_{lc}, see Pottmeier (2007). The following parameters are used throughout this contribution: $g_{safe} = 0$, $t_{lc} = 1$ and $\beta = 1$.

6.6 RESULTS

The simulations indicate that the two-lane model was capable of reproducing various features of real traffic. To prove this, a scenario of a two-lane section with an on-ramp was at first analyzed and the emerging dynamics were displayed. Just as in the single-lane model by Lee et al. (2004, 2005) the two-lane system showed the empirically found features of a system in the vicinity of an on-ramp. The various kinds of synchronized traffic could be reproduced in great detail, as could the pinch-effect.

The simulation scenario conserned a two-lane section of a highway with an on-ramp near the downstream end. At the on-ramp, the vehicles were inserted with constant frequency and time lag, simply corresponding to the reciprocal inflow rate. The inflow rates per time-step on the main road and on the on-ramp were denoted q_{main} and q_{ramp}, respectively. The total system length was set to 25,000 cells = 37,5 km. The relevant section of length $L = 10,000$ cells = 15 km including the on-ramp beginning at 13.5 km was cut out and displayed in space-time plots (Figs. 6.1–6.3). The length of the on-ramp was set to $L_{ramp} = 400$ cells = 0.6 km, which represented the standard length of a simple on-ramp in the highway network of North-Rhine Westphalia. The simulation time was 10,000 s ≈ 3.8 h after a relaxation period of the same duration.

The synchronized traffic was induced on the right lane and then passed over to the left lane. This was also the case for the emerging compact jam on the right lane

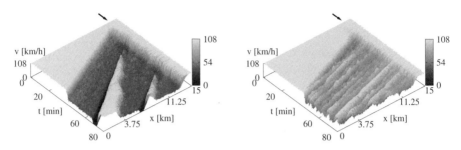

Fig. 6.1 The impact of an on-ramp on a two-lane road: The induced synchronized flow passed over from the right lane (right) to the left lane (left) ((q_{main}, q_{ramp}) = (0.56, 0.28)). The position of the on-ramp is indicated by the arrow.

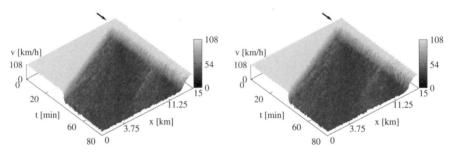

Fig. 6.2 A widening synchronized pattern ((q_{main}, q_{ramp}) = (0.56, 0.28)).

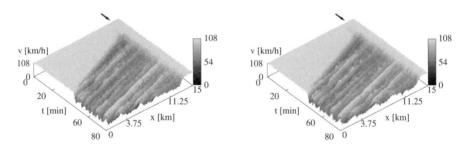

Fig. 6.3 A moving synchronized pattern ((q_{main}, q_{ramp}) = (0.58, 0.18)).

which migrated to the left lane. The widening and moving synchronized patterns were also reproduced (Figs. 6.2 and 6.3) as well as the localized synchronized or general pattern (Fig. 6.1). It should be noted that Figures 6.1 and 6.2 are results of the same inflow rates, and that the short arrow points at the position of the on-ramp.

Figure 6.4 presents the general appearances of the different traffic states collected in a diagram for which the composition mimics a phase diagram. The various regions were qualitatively different, which evoked the term "phase" in the description of the diagram. The x-axis displays the inflow at the on-ramp J_{ramp}, and the y-axis

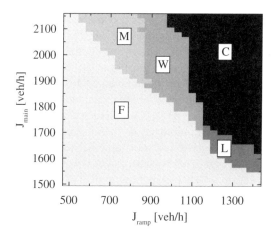

Fig. 6.4 The phase diagram of the open two-lane system with an on-ramp. The resulting regions stand for the different traffic states: "C" denotes the region of wide moving jams, "W" the widening synchronized traffic, "M" stands for moving synchronized traffic, "L" marks the region of localized synchronized traffic and "F" hints at the free-flow regime. All proposed variants of synchronized traffic were reproduced and were in good agreement with the empirical findings (Kerner, 2002).

depicts the flow J_{main} on the main road. The resulting areas represent the different traffic states: "C" denotes the region of compact jams, "W" the widening synchronized traffic, "M" stands for moving synchronized traffic, "L" marks the region of localized synchronized traffic, and "F" hints at the free-flow regime. The resulting phase diagram was, as stated elsewhere (Lee et al., 2004, 2005), comparable to those shown in Kerner (2002), Kerner and Klenov (2002) and Kerner (2003). All the proposed traffic patterns were reproduced in good accordance with the empirical findings.

One should note that the boundaries of the specific regions of the traffic states were not strict, however, both adjacent states could generally emerge at the limiting flow rates. The state that developed at the end of most simulations is shown in the diagram. This means that if a jam emerged, the state was classified as a general pattern. Otherwise, if synchronized traffic governed the system, the according classification of the synchronized pattern was applied. In the phase diagram, that pattern having occurred the most often is depicted. This is especially important for the boundary between widening and localized synchronized traffic, on the one hand, and the regime of the general pattern, on the other hand. At these boundaries, the traffic was not stable and whether the synchronized flow remained or the local perturbations became strong enough to form a seed for a wide moving jam was stochastic. Thus, it was worthwhile to analyze the life-time of the synchronized traffic with respect to the inflow rates, which is the topic of the next section. The analysis showed that the life-time varied significantly for the parameters of the boundary states and that the variance of the life-times was very high. For a detailed analysis of the life-times of the synchronized traffic and its application to ramp metering processes (see Pottmeier, 2007).

The time headway distribution $P(t_h)$ is here discussed as an example of the good accordance of the simulation results with empirical single-vehicle data,. The simulation results were calculated in a periodic two-lane system. Many headways lower than one second were found and the lower cut-off was at $t_h = 0.5$ s, which corresponded well with the empirical data. In both qualitative and quantitative terms, these results were in very good agreement with the empirical data (see Figs. 6.5 and 6.6).

The time headway distribution showed a much better correspondence with the empirical findings in the synchronized traffic than in the single lane system (see Lee et al., 2004). The empirically observed short time headways below one second were now present. As shown in Figure 6.5, time headways down to $t_h = 0.5$ s appeared, which corresponded to the cut-off time found in the empirical data.

It should be noted that the empirical data were also measured on a multi-lane road. Headways up to $t_h \approx 3$ s were found to occur in the empirical data as well as in the simulation results. Only the maximum was slightly shifted to higher times in the simulations.

The process of lane-changing seemed to play an important role for the occurrence of short time headways in the synchronized traffic. It was a significant result that the time headway distribution for the two-lane model matched the empirical data better than the single-lane model, especially in the synchronized traffic state.

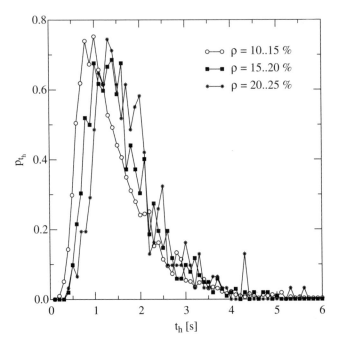

Fig. 6.5 Empirical time-headway distribution $P(t_h)$ in synchronized traffic. Short time-headways below one second were very frequent. For larger occupation ratios, the time-headway distribution was shifted to longer times. It should be noted that the distribution was normalized, i.e. $\Sigma P(t_h) \Delta t_h = 1$. From Knospe et al. (2002).

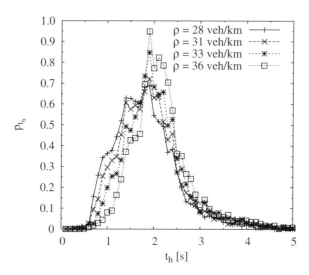

Fig. 6.6 The time headway distribution of the simulation in synchronized flow. Headways below one second were frequently detected. This was also found in the empirical data (Fig. 6.5). Furthermore, the curve was shifted to longer times for increasing densities. It should be noted that the densities given in this figure corresponded to the intervals used in Figure 6.5. For example, 28 veh/km corresponded to an occupation of 20% if σ_{max} = 140 veh/km (see e.g. Chowdhury et al., 2000 for the definition of the maximum occupation σ_{max}).

6.7   SUMMARY AND OUTLOOK

This contribution has presented an advanced cellular automaton model for two-lane traffic. The lane changing rule-set enhanced the cellular automaton model proposed by Lee et al. (2004, 2005) and this approach reproduced all empirically observed patterns of synchronized traffic as well as the pinch effect in the presence of an on-ramp. Moreover, it was in good agreement with the empirical data. This extended model thus allows for the simulation of complex traffic scenarios.

   Especially the influence of various kinds of disturbances or defects and its complex features in real world traffic could be qualitatively reproduced in good agreement by this model. Further, the proposed phase diagram was in accordance with the empirical data. Thus, this model approach offered the opportunity of obtaining a deeper insight into the dynamics of these disturbed systems. For example, the dynamic of this model demonstrated the randomness of jam generation in the vicinity of an on-ramp. This may be a new approach for analyzing and formulating ramp-metering algorithms since it takes into account not only the complex phases that emerge at an on-ramp but also the stochastic nature of the influence of on-ramp disturbances on the main carriageway. Preliminary results have shown that this model can reproduce the differences between a controlled inflow, where all vehicles are uniformly inserted, and an uncontrolled on-ramp, where the vehicles enter the main-carriageway with random time lags (Pottmeier, 2007). In the latter case, the probability of the emergence of a breakdown, corresponding to the empirical findings, is greater.

6.8 REFERENCES

Barlović, R., Huisinga, T., Schadschneider, A., and Schreckenberg, M. (2002) *Phys. Rev. E* 66:046113.

Barlović, R., Santen, L., Schadschneider, A., and Schreckenberg, M. (1998) *Eur. Phys. J. B* 5:793–800.

Barlović, R., Schadschneider, A., and Schreckenberg, M. (2001) *Physica A* 294:525–538.

Chowdhury, D., Santen, L., Schadschneider, A. (2000) *Phys. Rep.* 329:199.

Kerner, B. S. (1998) *Phys. Rev. Lett.* 81:3797.

Kerner, B. S. (2002) *Phys. Rev. E* 65:046138.

Kerner, B. S. (2003) In *Traffic and Granular Flow '01*, edited by M. Fukui, Y. Sugiyama, M. Schreckenberg, and D. E. Wolf, Tokyo, Springer, pp. 13–50.

Kerner, B. S., and Klenov, S. L. (2002) *J. Phys. A: Math. Gen.* 35:L31 (2002).

Kerner, B. S., Klenov, S. L., and Wolf, D. E. (2002) *J. Phys. A: Math. Gen.* 35:9971.

Knospe, W., Santen, L., Schadschneider, A., and Schreckenberg, M. (2000) *J. Phys. A* 33:L477.

Knospe, W., Santen, L., Schadschneider, A., and Schreckenberg, M. (2002) *Phys. Rev. E* 65:056133.

Knospe, W., Santen, L., Schadschneider, A., and Schreckenberg, M. (2004) *Phys. Rev. E* 70:016115.

Lee, H. K., Barlović, R., Schreckenberg, M., and Kim, D. (2004) *Phys. Rev. Lett.* 92:23702.

Lee, H. K., Barlović, R., Schreckenberg, M., and Kim, D. (2005) In *Traffic and Granular Flow '03*, edited by S. P. Hoogendoorn, S. Luding, P. H. L. Bovy, M. Schreckenberg, and D.E Wolf, Springer, Berlin.

Pottmeier, A. (2007) *Realistic Cellular Automaton Model for Synchronized Two-Lane Traffic*, Dissertation, URL: http://duepublico.uni-duisburg-essen.de/servlets/DocumentServlet?id = 17370, URN: urn:nbn: de:hbz:464-20080208-091511-8, University Duisburg-Essen (2007).

Pottmeier, A., Thiemann, C., Schadschneider, A., Schreckenberg, M. (2007) In *Traffic and Granular Flow '05*, edited by A. Schadschneider, T. Pöschel, R. Kühne, M. Schreckenberg, and D. E. Wolf, Springer, Berlin.

Sparmann, U. (1978) "Spurwechselvorgänge auf zweispurigen BAB-Richtungsfahrbahnen", in Forschung Straßenbau und Straßenverkehrstechnik Heft 263 (Bonn-Bad Godesberg, Bundesministerium für Verkehr).

Wagner, P., Nagel, K., and Wolf, D. E. (1997) *Physica A* 234:687.

PART III

People-Centered and Rail Simulation

CHAPTER 7

PEDESTRIAN SIMULATION TAKING INTO ACCOUNT STOCHASTIC ROUTE CHOICE AND MULTIDIRECTIONAL FLOW

Miho Asano, Masao Kuwahara, Agachai Sumalee

This chapter proposes a framework for a dynamic pedestrian model taking into account strategic route choice decisions and multi-directional flow propagation. In an open space, a large number of possible trajectories for pedestrians can be defined. However, psychologically, these trajectories are not necessarily considered as distinct routes in the route choice decision of the pedestrians. In the study described herein, pedestrians are assumed to choose a series of consecutive sub-areas to traverse along at first level. Subsequently, the model allocates the actual flows on each trajectory based up on an assumption of dynamic user optimum (DUO). By combining these two stages of flow allocation, the model assumes a hierarchical decision of a pedestrian's route choice. This chapter thus presents the adoption of a modified cell-transmission model (CTM) to represent the physical phenomena of congestion and dynamical propagation of pedestrian flows. A key difference between this model and the original CTM is the multi-directional congestion due to the nature of the pedestrian flows. An approach has thus been proposed to extend the analysis in the original CTM to consider multi-directional movement. The contribution also applies the proposed model to two test cases in order to illustrate its applicability and plausibility.

7.1 INTRODUCTION

Recently, the importance of walking as a part of a trip has regained a significant interest from transport planners. This is also supported by the realization of the important role of walking in a successful sustainable transport development. In addition, walking constitutes the major mode of transport in several large cities with very high densities of population and public transport services (e.g., in Tokyo, London, or Hong Kong). Nevertheless, there has been a lack of appropriate modeling tools to help plan an improvement in pedestrian infrastructure. The main research issues for pedestrian model development can be categorized into two main groups: (i) a plausible framework for pedestrian route choice or movement behavior and (ii) a realistic model for representing a physical congestion and interaction of pedestrian flows.

The space for physical movement of the pedestrian model can be considered as a continuum as compared to the discrete representation of a road network. Therefore, the concept of preferred routes or trajectories for pedestrians in this case differs considerably from those normally considered in the road network. In a continuum space, even after applying a form of discretization scheme, a potentially massive number of trajectories or paths for pedestrian movement can be defined. Hoogendoorn and Bovy (2004) proposed a continuum-space pedestrian model allowing the pedestrian to choose the activity location, schedule and route (defined as a trajectory in a continuum space). However, psychologically, these trajectories or paths are not considered as distinct options, nor are they chosen prior to the actual movement.

This chapter proposes a hierarchical route choice model in which, at the upper level, the pedestrian considers a route choice to consist of a series of areas to traverse. Subsequent to this upper level of decision, the pedestrian's actual movement or trajectory within each area is determined by the principle of Dynamic User Optimum, DUO (Kuwahara and Akamatsu, 1997). Similar to the position made by Hoogendoorn and Bovy (2004), the actual movement of the pedestrian is in this case divided into two stages, i.e., tactical (area choice) and operational (actual trajectory within each area) stages. Antonini et al. (2006) used an empirical data and discrete choice model to illustrate that, when possible people move toward pre-determined routes, which can be defined at a higher level of decision in contrast to the trajectories actually made at the operational level.

For the congestion and flow-propagation model, the nature of pedestrian flow and the flexibility of the pedestrian movement (i.e., without directional constraints of flow), requires a more complex representation in order to reproduce a realistic delay-density relationship. Several researchers have attempted to employ a micro-representation of pedestrian behavior, similar to the micro-simulation model for traffic (Helbing and Molnar, 1995; Blue and Adler, 2001). The other approach is to consider pedestrian flows following the development in the traffic area in which a form of aggregate speed-flow relationship can be adopted. Hughes (2002) developed a theoretical framework for bi-directional pedestrian speed-flow based on the fluid mechanics. Lam et al. (2003) proposed a speed-flow formulation for bi-directional traffic which was validated by the data from a study concerning various Hong Kong walkways.

The present chapter proposes an aggregated dynamic flow model for representing pedestrian congestion phenomena. Although aggregated pedestrian flow models, allowing the existence of more than one pedestrian in one cell, have been proposed by Naka (1978) as well as Teknomo and Millonig (2007), these models do not guarantee the relation to kinematic wave theory. The study reported herein, on the other hand, applies the Cell Transmission Model (CTM) proposed by Daganzo (1994, 1995). CTM is a macroscopic flow model equivalent to the kinematic wave theory for the vehicular traffic case. Nevertheless, for pedestrian flows, it is possible that more than one direction of flow interacts inside the space, especially in a crowded area (e.g., a subway station or a shopping center). As the original CTM does not handle a case with multi-directional flow/density, the present investigation has extended the formulation to consider the multi-directional flows for modeling the congestion from pedestrian flows.

This chapter is structured into five further sections. The next section gives an overview of the integrated route choice and dynamic pedestrian modeling framework proposed. Subsequently, the third section explains the concept of the route choice model developed herein, and includes a discussion on the tactical route choice model. The fourth section moves on to discuss the dynamic flow-propagation model in which an extension of the original CTM to the case of multi-directional flow is explained. The fifth section presents some tests with the route choice and flow-propagation model and the final section concludes the chapter and discusses future research.

7.2   FRAMEWORK OF AN INTEGRATED DYNAMIC PEDESTRIAN ROUTE CHOICE AND FLOW MODEL

The overall framework of the integrated dynamic pedestrian route choice and flow simulation model proposed in this chapter is shown in Figure 7.1. Similar to a standard vehicle simulation model, this model consists of three components: route generation, route choice, and flow propagation models. In this chapter, the terms route and path take on different meanings, as shown in Figure 7.2. A path refers to a trajectory

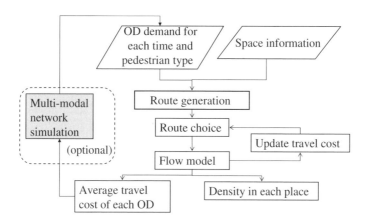

Fig. 7.1 The framework of a pedestrian simulation.

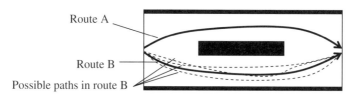

Fig. 7.2 The definition of route and path.

of pedestrians, whereas a route is a collection of paths sharing the same sequence of related areas. Although there potentially exists an infinite number of paths, pedestrians only perceive a limited set of routes defined by a series of consecutive sub-areas in the considered area. Therefore, the proposed model comprises a hierarchical model for trajectory determination. The upper level model aims at determining routes by using the sub-area-based networks, while the lower level is a path-choice model defining the exact trajectory of a pedestrian flow in the sub-areas.

A route-generation model defines a choice set of possible routes for each group of pedestrians. The model collates similar paths into a route based on the topology of the area, and the pedestrian then decides which route to take based upon the generalized travel cost on each route. The route choice model can rely on either deterministic (i.e., Wardrop's equilibrium) or stochastic user equilibrium (SUE) principles.

For each route, the demand associated with it is dynamically loaded onto several paths (or trajectories) by the flow propagation model. The model performs the dynamic flow propagation and calculates the delay that occurs at each point and time. The flow propagation model is based on the cell-transmission model (CTM) that divides a continuum space into a number of cells, and a pedestrian can move in and out of each cell with the delay based on the speed-density relationship considering the interaction of multi-directional flows.

7.3   NETWORK-BASED ROUTE CHOICE MODEL

7.3.1   Network structure for pedestrians

There are two types of space representations for the pedestrian route choice model: (i) a link-based model and (ii) a potential model (see Fig. 7.3 for graphical examples). The link-based model represents the topology of the area by a set of links and nodes in which an area can be defined by a node and a possible path can be defined as a link (see e.g., Gloor et al., 2004). With this representation, a number of existing theories and results from the vehicular network analysis field can be directly applied to the pedestrian model. However, due to a rather open space and free movement of pedestrians, the definition of appropriate relationship between walking space and links in the network can be problematic. The second form of space representation is the so called "potential model" (e.g., Hoogendoorn and Bovy, 2004). In this representation, an area is simply defined as a continuum plane in which pedestrian flows interact at each vector location inducing a form of delay. The pedestrian in this model then finds the shortest path from any possible continuous trajectories. As shown in Figure 7.3 below, a contour of cost (or delay) at each continuous location on this plane can be defined, in which the shortest path from one point to another is the trajectory with the steepest descent direction on this contour line. As mentioned, this model assumes that pedestrians choose a path from the infinite set of possible trajectories contradicting our preliminary observation with regard to the perception of possible "routes" in the space.

Although the number of possible routes in the continuum space is infinite, pedestrians in the actual situation are likely to perceive only a certain finite numbers

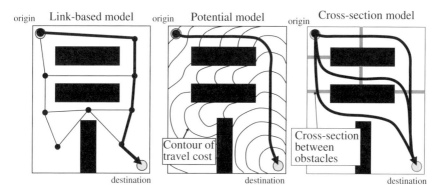

Fig. 7.3 The network structure for the route choice of a pedestrian.

of available trajectories. For instance, from the topology of the area in Figure 7.3, one can simply define three main possible routes from the origin to the destination, and not an infinite number of routes as defined in the potential model. Here, a model referred to as the "Cross-section model" is defined, introducing a number of intercept lines or cross-section lines between obstacles. The model distinguishes each of the areas, defined as units by a number of cross-section lines (without any other cross-section lines in the middle of the area), as sub-areas. In other words, the cross-section lines divide the whole walking space into a number of unique continuous walking sub-areas that are often surrounded by obstacles (as shown in Fig. 7.3). The pedestrians are then assumed to perceive a distinct alternative trajectory of journey (at the tactical level) by a series of consecutive cross-section lines (or sub-areas). Herein, the route is represented as a combination of these cross sections.

For generating sub-areas or cross sections in a common case, general rules or requirements must be defined. Two main requirements for a valid cross-section and sub-area include that (i) two different cross sections may not cross each other, and (ii) a sub-area consists of only walkable space defined by a finite set of cross-sections. By using various computational algorithms, cross-sections and sub-areas can be automatically or manually defined from a topology of an area. Figure 7.4 shows an example of the resulting cross-sections and sub-areas for an open urban space.

7.3.2 Route generation and route choice

After the definition of a number of valid cross sections and sub-areas, they are topologically replaced by nodes and links, whilst the connections between each cross section and sub-area are retained, as illustrated in Figure 7.5. This dual graph created by the process mentioned earlier still contains the key characteristic of the original representation in terms of the set of possible routes since the connectivity between the various sub-areas in an open area is maintained. Thus, the set of routes for pedestrians can still be generated by applying a form of K-shortest path algorithm (e.g., that of van der Zijpp and Catalano, 2005) to this dual graph as applied to the vehicular network.

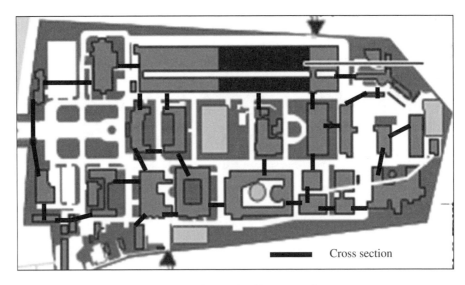

Fig. 7.4 An example of valid cross-sections defined in a general case.

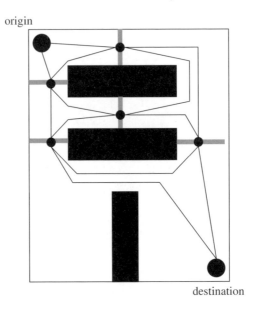

Fig. 7.5 A dual network of a cross-section-based network.

If $C_{ij}(t)$ represents the pedestrian travel cost between cross section i and j at time t (departure time from cross section i), then, obviously, depending on which positions on the cross sections i and j that are considered, the travel time differs. However, for an aggregated measure of travel cost between the cross section i and j at time t, the mean travel cost $E[C_{ij}(t)]$ between all points on each cross section can be adopted. The cost, $C_{ij}(t)$, (which may be assumed as simply the travel time) is defined by the flow propagation model.

Given a set of mean travel costs between all cross sections at all times in the study area, the costs of each feasible route, denoted $C_r(t)$, can be defined.

$$C_r(t) = \sum_i \sum_j \delta_{ij} E[C_{ij}(t)] \qquad (7.1)$$

Here, t is the departure time from its origin, δ_{ij} is 1 if and only if the path from cross section i to j is a part of route r and is otherwise 0. Since the possible choice sets of walking paths are aggregated based on cross sections, the route can be treated as a set of links in the dual graph and $E[C_{ij}(t)]$ is considered as the link travel cost of these links. This structure is similar to that of conventional traffic networks.

Due to this similarity, various tactical route choice models used in vehicle networks can be adopted based upon different definitions. Here, the concept of Stochastic Dynamic User Optimum (SDUO) was applied, in which the pedestrian with each origin-destination chooses the route based on a form of discrete choice model (e.g., logit or probit). In these discrete models, disutility can be defined for each route incorporating $C_r(t)$ in addition to other factors, including avoidance to some sub-areas or particular physical barriers.

7.4 PATH CHOICE MODEL

In the previous section, the tactical route choice was explained with $E[C_{ij}(t)]$. This expected travel time is based on the result of flow propagation inside the sub-area. In this stage, a detailed direction of the flow propagation should also be considered. The sub-areas are discretized into a number of square cells, 2–3 m in length, as shown in Figure 7.6. Adjacent cells are connected to each other by tangential surfaces. Similarly

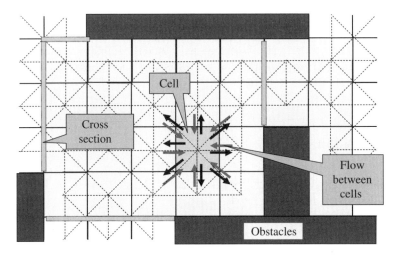

Fig. 7.6 A cell-based network.

to sub-area-based networks in the upper level, a cell-based network is structured using connection links between the cells. Since the Cell-Transmission model is applied to the cells, as described later, there is an assumption that the flow-density condition in one cell is uniform and this determines the size of the cells. If the cell size is too large, a pedestrian may not be uniformly distributed in the cell and it becomes problematic to represent the condition of a cell by a single flow-density curve. If, on the other hand, the cell is too small, i.e., comparable to the size of the pedestrians, it may again be difficult to draw the flow-density based on information of maximum one person in the cell.

Given that a pedestrian traverses from cross section i to j (defined by the tactical route choice), the actual trajectory or the operational level of path choice principally depends on the concept of dynamic user optimum (DUO) in which the pedestrian traverses along the shortest path, considering the path cost at departure time t from the cross section i.

The travel time from one cell to another is measured as a result of the flow propagation model. If no pedestrians move between the cells during the measurement time, the travel time $TT_{pq}(t)$ from cell p to q at time t is determined using the density of the destination cell q and the speed-density relationship given in the CTM.

$$TT_{pq}(t) = \frac{L_{pq}}{v_q(t)} \tag{7.2}$$

Here, L_{pq} is the link length or distance between the centers of cells p and q, and $v_q(t)$ is speed of a pedestrian in cell q at time t. The pedestrians in this sub-area moving from cross section i to j are assigned based on the DUO principal.

$E[C_{ij}(t)]$, i.e., the expected travel time between cross sections, corresponds to the calculation result of path choice and flow propagation. However, in order to simplify the calculation, $E[C_{ij}(t)]$ can be approximated to the minimum travel cost between i and j. This is because the travel time of all the pedestrians will be the same if the flow is under a dynamic user equilibrium (DUE) in which pedestrians traverse along the shortest path with a given path cost throughout the entire trip. Although DUE and DUO do not provide exactly the same travel cost information from current time to the end of the trip, the trend of the travel cost under these principals should show close similarities.

7.5   FLOW PROPAGATION MODEL

As discussed in the previous section, pedestrians with a given route are loaded to the network and a flow propagation model is applied in order to represent the congestion phenomena. Both microscopic and macroscopic flow propagation models could be applied in this framework, and analysis was recently performed from both points of view by using experimental as well as observed data. This chapter introduces a macroscopic model based on the cell transmission model.

7.5.1 Cell Transmission Model

The space in the sub-area is discretized into a number of cells and a macroscopic flow-density curve, representing the dynamical movement of the pedestrians, is applied to each cell.

The concept of the cell transmission model (CTM) as proposed by Daganzo (1994, 1995) is here utilized to define a plausible CTM for pedestrian flows. In CTM, the number of vehicles, $y_i(t + \Delta t)$, moving from cell $i - 1$ to i at time $t + \Delta t$ can be calculated considering the possible number of vehicles that can exit cell $i - 1$, $P_{i-1}(t)$. Moreover, the possible number of vehicles that can enter cell i in Δt, $S_i(t)$. $P_{i-1}(t)$ and $S_i(t)$ is obtained according to the flow density curve given in Figure 7.7. $Q_i(t)$ is the maximum number of vehicles that can move from/to cell i to/from the other cells from time t to $t + \Delta t$.

$$P_{i-1}(t) = \min\left\{n_{i-1}(t), Q_{i-1}(t)\right\} \qquad (7.3)$$

$$S_i(t) = \min\left\{Q_i(t), \frac{w}{v_f}(N_i(t) - n_i(t))\right\} \qquad (7.4)$$

Here, $n_i(t)$ is the number of vehicles in cell i at time t, and $y_i(t)$ is calculated by taking the minimum of these two values.

$$y_i(t) = \min\left\{P_{i-1}(t), S_i(t)\right\} \qquad (7.5)$$

The original CTM has a restriction with regard to the setting of the size of a cell, L, and the simulation time step, Δt, to guarantee its equivalence to the kinematic wave theory defined as:

$$v_f \cdot \Delta t = L \qquad (7.6)$$

Three problems occur when applying this original CTM to the pedestrian flow model. The first is that the walking distance of the diagonal flow is longer than that of the horizontal and vertical flows. This property does not satisfy the requirement in

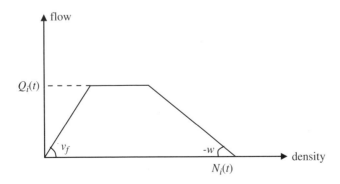

Fig. 7.7 The flow-density relationship used for CTM.

Eq. (7.6) guaranteeing its consistency with the kinematic wave theory. The second problem is the effect of multi-directional flows in a single cell. In the original CTM, the model is defined for a case with uni-directional flows, especially for the delay calculation and dynamic flow propagation. However, in the case of a pedestrian model, there are eight possible directions of flow movements from one cell to other adjacent cells. Various compositions of flows from a variety of directions may result in a rather different level of delay in that cell. In addition, the level of delays experienced by the pedestrians in various directions, even located within the same cell, may also differ (similarly to the case of asymmetric delay functions in a traffic network). The final issue is related to merging and diverging behaviors of pedestrians, which can take place in any cell. The next section discusses some modifications to the original CTM with the objective of overcoming these issues.

7.5.2 Modification of CTM

(1) *Simulation time step.* Under the assumption that flows of different direction j occur in a cell; if v_f is equivalent for all directional flows, the simulation time step, Δt_j must satisfy the following condition:

$$\Delta t_j = \frac{L_j}{v_f} \quad \text{where} \quad L_j = \begin{cases} L & \text{if } j \text{ is horizontal/vertical flow} \\ \sqrt{2}L & \text{if } j \text{ is diagonal flow} \end{cases} \tag{7.7}$$

To overcome this problem, the simulation time interval Δs is set to a smaller value than that of Δt_j, such that

$$\Delta t_j = m_j \Delta s \quad m_j : \text{integer} \tag{7.8}$$

The simulation model then calculates the flow propagation for each direction j with an interval of Δt_j.

(2) *Effect of multi-directional flow.* The CTM uses a speed-density curve in each cell to determine the actual amount of flow moving from one cell to another. However, the speed of a pedestrian in the multi-directional flow case decreases due to the conflict between pedestrians with different speeds and directions. The speed of a pedestrian depends not only on the number of pedestrians in the cell, but also on the desired direction and speeds of the other pedestrians. In order to calculate the speed of a pedestrian in a certain position in the cell, the concept of converted density is adopted in which a conversion function is applied to the mixture of multi-direction flows so as to define an equivalent density for that cell under the uni-directional flow case.

The converting function is based on the densities of pedestrians in different directions in the cell. Depending on the condition of the case in question, a variety of converting functions can be defined. For example, under highly congested conditions, Algadhi and Mahmassani (1990) proposed a function of the speed-flow relationship for the orthogonal crossing situation based on empirical data. On the other hand, during free flow or very mild congestion conditions,

Naka (1978) concluded that the effect of crossing flows on the speed reduction was insignificant.

An example of the converting function is shown below, using the speed-density function of Algadghi and Mahamassani. Their speed-density function for opposing flow is:

$$v_1 = 0.58 \cdot \left(1 - \frac{k_1 + 0.56k_2}{k_{jam}} \right) \qquad (7.9)$$

where v_1 is the speed of direction 1 (m/s), k_1 and k_2 are the number of pedestrians going in direction 1 and 2, respectively, and k_{jam} is the jam density estimated as 5.85 ped/m². In this case, the impedance of k_2 on v_1 is 0.56 times larger than that of k_1. Therefore, $F(n_2, 1) = 0.56\, n_2$ would be the converting function in this case.

(3) *Merging and diverging.* The original CTM considers the merging and diverging behaviors separately at intersection cells. However, in the pedestrian model, merging and diverging can take place in any cell since pedestrians are free to move in all directions. Considering the flow from cell i to cell j: cell i has adjacent cells $r(\in \Omega$: set of adjacent cells) and cell j has adjacent cells $s(\in \Phi$: set of adjacent cells).

In the case of a multi-directional flow, the total outflow from cell i, i.e. $\sum_r P_{ir}(t + \Delta t_{ir})$, is assumed to be distributed proportionally to all possible outflows in all directions, $P_{ir}(t + \Delta t_{ir})$. Thus, from Eq. (7.3), the possible outflow from i to j can be written with the density converting function as:

$$P_{ij}(t + \Delta t_{ij}) = \min\{F(n_{ij}(t), j), \frac{F(n_{ij}(t), j)}{\sum_r F(n_{ir}(t), j)} Q_i(t)\} \qquad (7.10)$$

where $F(n_\alpha, \beta)$ denotes the converted density. For simplicity, $F(n_{ij}(t), j)$ is referred to as $n_{ij}(t)$ in the rest of this chapter. The total number of possible pedestrians who can enter cell j from all adjacent cells s can be defined as:

$$\sum_s S_{sj}(t + \Delta t_{sj}) = \min\{Q_j(t), \frac{w}{v_f}(N_{jam}(t) - \sum_s n_{sj}(t))\} \qquad (7.11)$$

The possible inflow from each direction is given based on a certain merging ratio. This merging ratio $R_{ij}(t + \Delta t_{ij})$ is assumed to be defined by the proportion of the demand from each direction to cell j.

$$R_{ij}(t + \Delta t_{ij}) = \frac{\min(P_{ij}(t + \Delta t_{ij}), S_{ij}(t + \Delta t_{ij}))}{\sum_s \min(P_{ir}(t + \Delta t_{ir}), S_{sj}(t + \Delta t_{sj}))} \qquad (7.12)$$

If the total possible outflow from all of the adjacent cells s to cell j is lower than the possible inflow from all s to j, then all of the flows can enter cell j. Otherwise, the amount of flows entering from each cell is distributed by $R_{ij}(t + \Delta t_{ij})$. Therefore, the actual number of pedestrians moving from cell i to j can be defined as:

$$y_{ij}(t + \Delta t_{ij}) = \begin{cases} P_{ij}(t + \Delta t_{ij}) & \text{if } \sum_s P_{sj}(t + \Delta t_{sj}) \le \sum_s S_{sj}(t + \Delta t_{sj}) \\ R_{ij}(t + \Delta t_{ij}) \sum_s S_{sj}(t + \Delta t_{sj}), & \text{otherwise} \end{cases} \quad (7.13)$$

7.6.   APPLICATION TO SIMPLE CASE STUDIES

7.6.1   Test for flow propagation model

The proposed flow propagation model was tested in two simple cases. The first study network was, as shown in Figure 7.8, designed to test the crossings of bidirectional flow with an angle of 90 degrees, whereas the other case examined the flow propagation while taking route choice into account. In each test, the flow-density relationship, according to Figure 7.9, was used and no multi-directional effects were considered, i.e. $F(n_\alpha(t), \beta) = n_\alpha(t)$. In the following applications, the simulation time step, Δs, was set as 0.1 s, and the size of the cells was 2×2 m². The direction of the flows was pre-determined and did not change during the simulation period since the sole purpose of these tests was to illustrate the congestion phenomena occurring in the modified CTM. The demand loaded to the study network 1 is represented in Figure 7.10, portraying the time-dependent demand loaded to each origin cell each second.

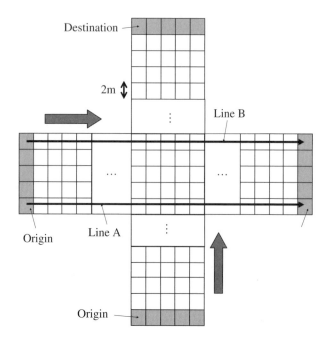

Fig. 7.8 Simulated network 1 with a 90-degree crossing.

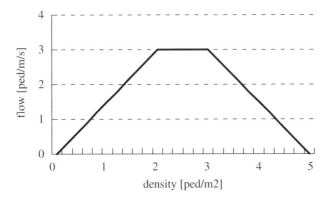

Fig. 7.9 The flow-density curve used in the simulation tests.

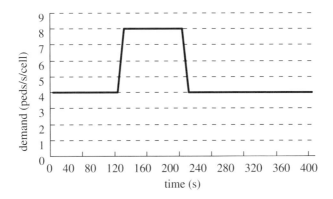

Fig. 7.10 The demands loaded to the simulation tests.

Figure 7.11 shows the evolution of pedestrian flows at various time steps. The value in each cell corresponds to the number of pedestrians in that cell at each time epoch. From this case, the congestion was mainly concentrated in the south-west corner of the cell, where the flows from a variety of directions merged. On the northern or eastern sides of the crossing area, the density of flows was much lower than in the south-western part. This was due to the flow entering the northern or eastern cells already having passed and been constrained by the capacities of the cells in the south-western side.

Two trajectories, Lines A and B, as shown in Figure 7.8, were investigated further, and Figure 7.12 presents the time-space diagrams for these lines. The shockwave phenomenon, which is similar to that of the vehicle flow, was observed in both cases. The queue vanishing-time for Line A was much longer than that of Line B. Since the flows of Line A merged directly with the orthogonal flows, the merging effect was much stronger than that of Line B. The amount of orthogonal flow was first constrained by the capacities of the cells on Line A, and thus the inflow of the orthogonal flow to cells on Line B were reduced. Hence, the pedestrian flows on Line B could pass the crossing section more smoothly.

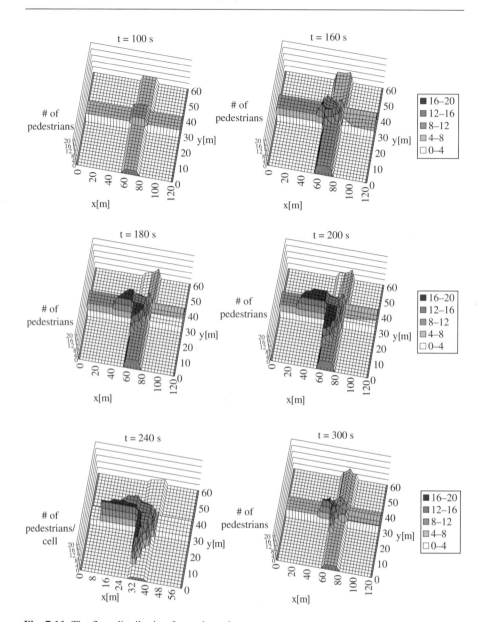

Fig. 7.11 The flow distribution for various times.

7.6.2 Network test

Figure 7.13 shows the second test problem, involving the route choice model. As can be seen, there was only a single origin and destination area, as well as one obstacle (dark color) in the middle of the figure. Two possible routes existed for traveling from the origin to the destination area. The first route was through the narrow corridor on the upper side of the obstacle, and the second route involved bypassing the obstacle on

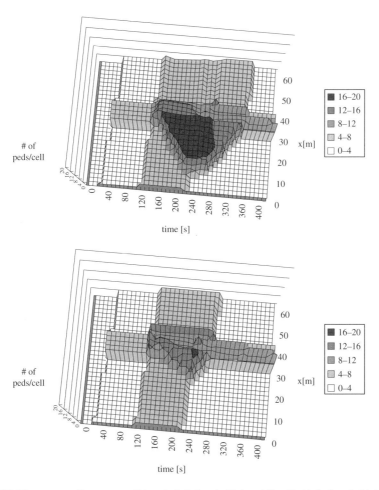

Fig. 7.12 Time-space diagrams on Line A (left-hand side) and line B (right-hand side) in the case of an orthogonal crossing.

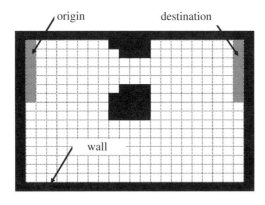

Fig. 7.13 Test network 2 for route choice (2×2 m^2 grids).

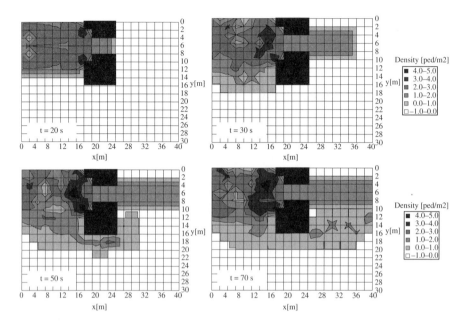

Fig. 7.14 The density distribution in network 2.

its lower side. A constant inflow demand profile with 12 people/sec was assumed, and the flow-density relationship was defined in the same way as in the previous test.

Figure 7.14 shows the pedestrian density distribution from the test. At the beginning, all the pedestrians preferred to traverse through the upper route ($t = 20$). After congestion occurred in this bottleneck, some of the pedestrians changed their routes to the second route (lower one). The density distribution in the congested area was unevenly spread, which could be the result of the update interval of the travel cost for the CTM and path choice models. However, the expansion of the pedestrian queue could be observed macroscopically.

In this model, the pedestrians do not diffuse if their density is below a critical value (= free flow condition), as shown in the area to the left of the obstacle in the figure at $t = 30$. Realistically, the actual pedestrian flow should spread over a wider area even in the free flow condition due to several reasons. Although there were no critical phenomena for the bottleneck evaluation, as illustrated in this example, it is possible to introduce additional factors into the model to calibrate the flow distribution in order for it to be more realistic.

7.7   CONCLUSIONS

This chapter has proposed an integrated framework of a dynamic pedestrian model with route choice decision. The key attributes of this model include the representation of a collection of trajectories in an open space as possible routes in the area in question. This could be done by using the concept of cross section lines defined by adjacent obstacles.

With this representation, the pedestrian chooses his/her route at the tactical level, which leads to the choice of a series of consecutive sub-areas as the main movement direction. At the operational level, the pedestrian then traverses through the sub-area by taking the shortest path based upon the Dynamic User Optimum principle. On the physical side of the model, the cell-transmission model (CTM) was extended to cope with a case of multi-directional flow. The route choice model was also validated in a simple network.

Future research will involve an evaluation of the entire modeling framework as well as a comparison against empirical behavioral data, particularly on the density-converting function and the flow propagation model. The connection between the route and path choice models will also be further investigated.

7.8   ACKNOWLEDGEMENTS

This research was funded by the Japan Society for the Promotion of Science.

7.9   REFERENCES

Algadhi, S. A. H., and Mahmassani, H. S. (1990) "Modelling crowd behavior and movement: application to Makkah pilgrimage", *Transportation and Traffic Theory*, 11:59–78.

Antonini, G., Bierlaire, M., and Weber, M. (2006) "Discrete choice models of pedestrian walking behavior", *Transportation Research Part B*, 40:667–687.

Blue, V. J., and Adler, J. L. (2001) "Cellular automata microsimulation for modeling bi-directional pedestrian walkways", *Transportation Research Part B*, 35:293–312.

Daganzo, C. F. (1994) "The cell transmission model: a simple dynamic representation of highway traffic consistent with the hydrodynamic theory", *Transportation Research Part B*, 28:269–287.

Daganzo, C. F. (1995) "The cell transmission model, part II: network traffic", *Transportation Research Part B*, 29:79–93.

Gloor, C., Stucki, P., and Nagel, K. (2004) "Hybrid techniques for pedestrian simulations", *In Proceedings of the 4th Swiss Transport Research Conference.*

Helbing, D., and Molnar, P. (1995) "Social force model for pedestrian dynamics", *Physical Review*, 51:4282–4286.

Hoogendoorn, S. P., and Bovy, P. H. L. (2004) "Pedestrian route-choice and activity scheduling theory and models", *Transportation Research Part B: Methodological*, 38:169.

Hughes, R. L. (2002) "A continuum theory for the flow of pedestrians", *Transportation Research Part B*, 36:507–535.

Kuwahara, M., and Akamatsu, T. (1997) "Decomposition of the dynamic assignments with queues: DUO and DUE", *Transportation Research Part B*, 31:1–10.

Lam, W. H. K., Lee, J. Y. S., Chan, K. S. and Goh, P. K. (2003) "A generalised function for modeling bi-directional flow effects on indoor walkways in Hong Kong", *Transportation Research Part A*, 37:789–810

Naka, Y. (1978) *Study on Complicated Passenger Flow in a Railway Station*, 1079 (in Japanese).

Teknomo, K., and Millonig, A, (2007) "A navigation algorithm for pedestrian simulation in dynamic environments", *Proceedings of the 11th World Conference on Transport Research.*

van der Zijpp, N. J. and Fiorenzo Catalano, S. (2005) "Path enumeration by finding the constrained K-shortest paths", *Transportation Research Part B*, 39:545–563.

CHAPTER 8

F.A.S.T. – FLOOR FIELD AND AGENT BASED SIMULATION TOOL

Tobias Kretz, Michael Schreckenberg

This chapter presents a floor field and agent based model (F.A.S.T.) of pedestrian motion. In an application, its parameters were fitted to one run in an evacuation exercise at a primary school. The simulations with these parameters were then compared to further runs during the same exercise.

8.1   INTRODUCTION

Understanding the dynamics of crowds in different situations has gained an increasing interest over the last decades (Schadschneider et al., 2009). Anyone who carefully watches the international media will note there occurs, world-wide, roughly one incident per month in junction with crowds with disastrous or almost-disastrous results (Helbing et al., 2002; Kretz, 2007).

Based on this background, it is not surprising that almost as long as it has been recognized that large crowds and people in large buildings may present some special problems, calculation frameworks have been set up to estimate evacuation times and other crucial values of pedestrian motion. Some of these have reached the status of off-the-shelf software products (ASERI, 2008; Klüpfel and Meyer-König, 2002; PTV, 2008; Thompson and Marchant, 1994).

The complexity of the models has increased with available calculational power. The road has gone from manual calculation (Predtetschenski and Milinski, 1971) and hydrodynamic models, through network models, to models with an individual representation (agents, "microscopic simulation") of real persons and an ever more exact inclusion of the environment, sometimes with discrete (Blue and Adler, 2000; Klüpfel, 2003; Nishinari et al., 2004), and sometimes with continuous (Helbing and Molnar, 1995) representations of space and time.

8.2   A MODEL OF PEDESTRIAN MOTION

This section presents a model of pedestrian motion, implemented into F.A.S.T. (*Floor field and Agent based Simulation Tool*). It is microscopic and makes use, to a large extent, of so called floor fields (Schadschneider, 2002) to determine the motion of the agents. The model is discrete in space and time, and the agents move on a grid of cells representing squares of 40×40 cm². The time advances in rounds and each round is interpreted as one second. Every cell can maximally be occupied by a single agent, and a new feature of this model as compared to similar preceding ones is the possibility to simulate agents with a speed larger than one cell per round.

The model makes use of floor fields (Burstedde et al., 2001; Nishinari et al., 2004; Schadschneider, 2002). Technically speaking, floor fields are look-up tables located in space, and their usage can be compared to that of the concept of a potential in physics. A grid is spread over space and each grid cell of the floor field contains a – scalar or vector – value. The purpose of this method is to transform long-range interaction into short-range ones, thereby removing the need to scan and consider a wider environment.

Floor fields fulfil two tasks:

- Constant floor fields, in a simple way, render it possible to save calculation time, since important values such as the distance of a cell to an exit are stored in them.

- Floor fields that change with time can be used to transform long-range interactions into short-range ones. This is a more sophisticated element in order to save calculation time.

The F.A.S.T. model consists of three floor fields:

- For each cell, the "Static floor field" (Burstedde et al., 2001; Nishinari et al., 2004) contains the information of the distance to the exit, and there is, in fact, one static floor field for each exit. (Exit-cells that are connected by a common edge are grouped to exits.) The static floor field can be understood as a kind of potential in which the agents "fall" towards the exit.

- The "Dynamic floor field" (Burstedde et al., 2001; Nishinari et al., 2004) is a vector field. An agent who has moved from cell (a, b) to cell (x, y) changes the dynamic floor field (D_x, D_y) at (a, b) by $(x-a, y-b)$ after all agents have moved. The dynamic floor field does not change on intermediate cells that the agents cross on their way from their source to their destination cell. Directly after this step, all values of both components of D decay with probability δ and diffuse with probability m to one of the (von Neumann) neighbouring cells. Since the vector components can be negative, the term decay signifies a reduction of the absolute value. Diffusion is only possible from x- to x- and from y- to y-components. Diffusion from a component with a negative value leads to a lowering of the component value at the target cell, whether it is positive or negative and vice versa for positive values.

- The distance of a cell to the next wall is also saved in a floor field if this distance is smaller than a certain threshold value. This is a simple construction that makes it possible to avoid calculating the distance every time it is needed.

There are other influences on the motion that are not governed by floor fields but rather by the properties of the agents:

- An inertia helps the agents to avoid sharp turns at high velocities. This inertia is not the normal inertia of Newtonian physics, but rather, due to the special construction of the human movement apparatus. A person can relatively easily accelerate or decelerate into the normal direction of motion, whereas significantly deviating from that direction on short time-scales is more difficult, especially at high velocities.

- When possible, moving too close to other people, is typically avoided and thus an agent can also have a repulsive effect on other agents. This is true provided that enough cells are available at a far enough distance to the agents.

- Some kind of "friction" (Kirchner et al., 2003) is implemented which reduces the effectiveness with which agents competing for the same cell during one round reach that cell.

- If the scenario includes more than one exit, the agents choose one of them at the beginning of each round. Here, the decision of the last round plays an important role, since once a decision has been made, it typically is not revised every second.

Except for the last two influences, the strength of all influences is determined by coupling constants: An agent "couples" to the static floor field, to the dynamic floor field, to the wall-field, to his own inertia and to the presence of other agents. All of these coupling constants can be interpreted in some way; the coupling to the static floor field, for instance, corresponds to the knowledge that the agent has of his environment, the coupling to his own inertia represents the ratio of strength to body mass or more generally his fitness.

Figure 8.1 shows how these influences are merged into a three-phase process in each round. At first, all agents choose the exit they want to approach during that particular round, according to the influences described above and to Eq. (8.1). In the second phase, all parallel agents choose a destination cell out of all cells that they can theoretically reach during that round (Kretz and Schreckenberg, 2006). Walls and cells that are occupied by other agents are excluded. The set of reachable cells is determined by the individual maximal speed of an agent. For each reachable cell and for each influencing effect, a value is calculated. Together with the corresponding coupling constant, a partial probability is computed according to Eqs. (8.2)–(8.6). Following Eq. (8.1), these are finally merged into a single probability for a reachable cell to be chosen as the destination cell.

In the last phase, all agents move and try to reach their destination cell. However, this is not necessarily accomplished if other agents intercept the path to that destination cell.

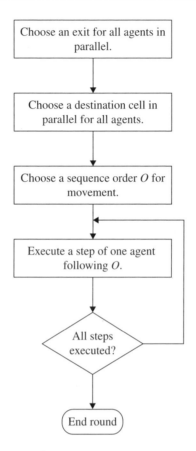

Fig. 8.1 The structure of one round.

Up until now, F.A.S.T. has been validated in a number of simple scenarios (Rogsch, 2005), and results of an evacuation exercise at a primary school have been compared to the results of a F.A.S.T. simulation of this exercise. Concerning the simple scenarios, F.A.S.T. provided results that were comparable to a still widely-used manual calculation method (Predtetschenski and Milinski, 1971). This implies that the evacuation times predicted by F.A.S.T. were typically more conservative (see Rogsch, 2005) than those predicted by some commercially available software. Additionally, the F.A.S.T. model has been used as an example to study oscillations in narrow bottlenecks (Kretz et al., 2006). Out of these examples, the following section presents a detailed report on the evacuation exercise at a primary school.

8.3   EVACUATION EXERCISE IN A PRIMARY SCHOOL

The evacuation exercise that was reported in (Klüpfel et al., 2002) was repeated, however with fewer pupils. The children were highly motivated, which was partly due to the presence of a camera team reporting for a children's news program of a German children's TV station. The exercise was repeated twice. The first time around, the music class did not become aware of the alarm due to the bell in their class-room being broken and the fact that they were singing so loud that they could not hear the bell from the floor. The school comprised two buildings: The main building and a newer second building. The music class was on the second of three floors in the main building. As for the second building, it had two floors, see Figure 8.2.

In addition to the video surveillance by the three cameras, the time was measured for the last person to leave the third floor. A person was considered to have exited the main building when reaching the last of the outside stairs.

8.3.1 Results

The results (Table 8.1) of the two exercises in the main building are not comparable since the music class only took part in the second exercise. The data of the second

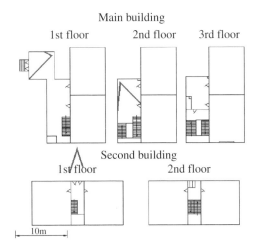

Fig. 8.2 The floor plan (American floor numbering). The red angles give the position of the cameras.

Table 8.1 Results.

Time (in seconds) after alarm for the...	First exercise	Second exercise
...last person to leave the main building	65.4	69.9
...first person to leave the main building	28.5	12.3
...last person to leave the 2nd floor of the main b	43.2	44.9
...first person to leave the 2nd floor of the main b	15.3	13.2
...last person to leave the 3rd floor of the main b	25.0	24.0
...last person to leave the second building	60.5	56.5
...first person to leave the second building	16.2	5.2

building, however, suggested that the students reacted faster to the alarm during the second run. Thus, either a learning effect had occurred, or the pupils – at least some of them – anticipated the alarm, instead of being surprised by it, see Figures 8.3 and 8.5.

8.3.2   Comparison to simulation results

After the exercise was finished and the empirical data was evaluated, simulations were carried out with the aim of obtaining the best possible reproduction of the empirical data of the first exercise – tuning the parameters by hand for this aim. This resulted in the evacuation graphs of Figure 8.4. Due to technical reasons, an empirical evacuation graph at the main exit could not be evaluated, however the total time – averaged over 1000 simulation runs – of the evacuation (until all pupils had completely left the main building) was 62.2 seconds with a standard deviation of 1.3 seconds. The smallest evacuation time that appeared during these 1000 simulation runs was 58 seconds, the largest 69 seconds. For the evacuation of the second floor the corresponding values

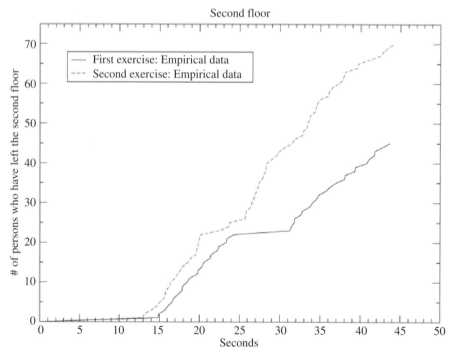

Fig. 8.3 Evacuation graphs of both runs of the second floor of the main building. The pupils were counted when they left the second floor and moved down the first step of the stairway between the second and first floors. In the first run (without the music class) the pupils of the other classes on the second floor left it just before the pupils of the third floor arrived. In the second run, the pupils from the third floor arrived before all the pupils of the second floor had left this floor. However, two pupils seemed to dawdle for no apparent reason, and thus another – but in this case smaller – plateau appeared in the evacuation graph.

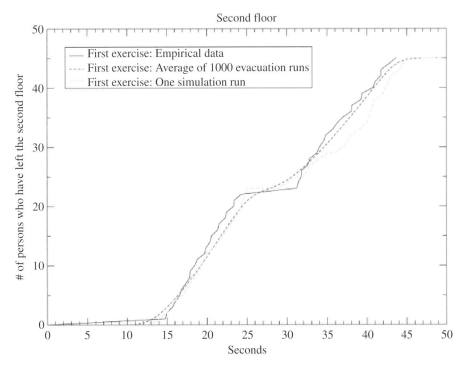

Fig. 8.4 A comparison of empirical and simulation results. The simulations were carried out after the exercise, making them a calibration of the simulation rather than a prediction.

were: 44.6 ± 1.4 seconds, with all evacuation times comprised between 42 and 52 seconds.

The parameters that gave rise to the results in Figure 8.5 were the following (see Eqs. (8.2)–(8.7) in Appendix A for an explanation of the parameters): $k_S = 3.0$, $k_D = 2.0$, $k_I = 2.0$, $k_W = 0$, trace strength: 6, $\alpha = 0.8$, $\delta = 0.5$, $\mu = k_P = 0$. For the reaction times of the teachers and the pupils on the third floor (fourth grade, oldest pupils of the school), the following distribution was used: $t_r^{min} = 18$ seconds, $t_r^{av} = 19$ seconds, $t_r^{max} = 20$ seconds, $t_r^{std} = 1$ second. The maximum speed was set to $v_{max} = 5$ (cells per round) for all pupils, while for the younger pupils the reaction time was set to smaller values in some cases: $t_r^{min} = 10$ seconds, $t_r^{av} = 15$ seconds, $t_r^{max} = 20$ seconds, $t_r^{std} = 5$ seconds and the speed varied: $v_{max}^{min} = 4$, $v_{max}^{av} = 6$, $v_{max}^{max} = 8$, $v_{max}^{std} = 1$. This corresponds to the following observations: Certain of the younger pupils were highly motivated, and speeds up to 3 m/s were observed. The older students of the third floor remained closer together and appeared to be slightly less (but still highly) motivated and/or more disciplined. It might come as a surprise that all pupils seemed to have such a strong inertia, but k_I should always be set and seen in relation to k_S and it was indeed a fact that the turns in the stairway significantly slowed the pupils down. Note: Even small variations in some parameters, such as the maximum speed, the reaction times, α, δ, the trace strength, k_D, k_I and to some extent k_S, led to a much smaller agreement between observation and simulation. It was especially difficult to find parameters that could reproduce the plateau in the evacuation graph.

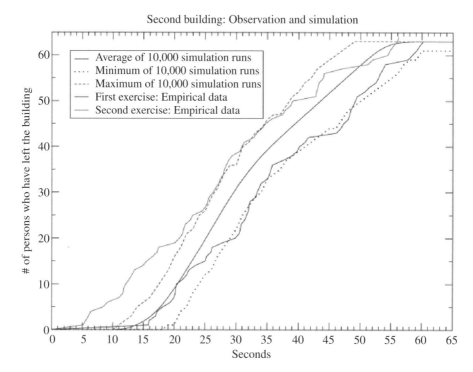

Fig. 8.5 A comparison of empirical and simulation results for the second building. During the second exercise, there was an additional person in the building. The simulation was carried out with the parameters that had been fitted to the results of the first run in the main building.

Subsequently, these parameters were used in simulations that included the music class (second run), and Figure 8.6 displays a comparison of the observed and simulated runs. The total time – averaged over 1000 simulation runs – of the evacuation was 67.7 seconds with a standard deviation of 1.7 seconds. The smallest evacuation time that appeared during these 1000 simulation runs was 63 seconds, the largest 75 seconds. For the evacuation of the second floor, the corresponding values were: 46.0 ± 1.7 seconds with all evacuation times comprised between 42 and 56 seconds.

While the parameters had been calibrated with the data of the evacuation of the second floor during the first exercise, the results of the simulation for the evacuation of the whole building during the second exercise (i.e., 67.7 ± 1.7 seconds, minimum 63, maximum 75 seconds) were in good agreement with the corresponding empirical data (69.9 seconds). The fact that no set of parameters could be found that fully reproduced the high outflow from the second floor was probably due to the smaller size of children as compared to adults, for whom data is normally taken in experiments and observations.

When applied to the second building, the same parameters led to an average simulated evacuation time of 56.0 ± 2.2 seconds, see Figure 8.5. As compared to the first exercise, the students in the second building performed better throughout the whole second exercise. It is not clear whether this was due to them anticipating

Second floor

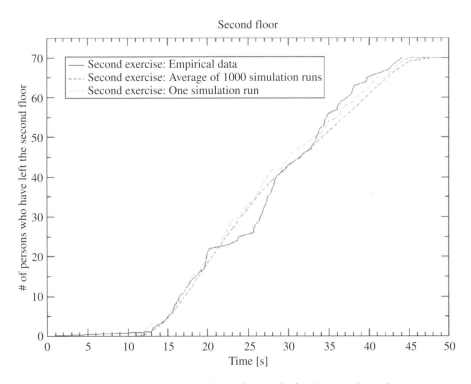

Fig. 8.6 A comparison of empirical and simulation results for the second exercise.

a second alarm or if they actually performed better during egress, due to a learning effect. The averages of the simulated evacuation times yielded results that were almost identical to those of the first exercise at the beginning of the process and that fit very well to the data of the second exercise at the end of the process.

8.4   SUMMARY

This work has presented a model of pedestrian motion. As an example of parameter gauging, results of the model were compared to an evacuation exercise, carried out in two runs, in a primary school. Within the variations between the two exercises, the results of the simulation were in good agreement with the empirical data. However, the agreement was better for identical floor plans and different population numbers as opposed to for identical population numbers and different floor plans.

The ability to simulate pedestrians with speeds larger than one cell per round and the possibility to distribute various speeds over the crowd were two main reasons for the close resemblance of the simulation results to reality. Furthermore, during the tuning of the parameters, the inertia (parameter k_I) played a crucial role.

It should be pointed out, that the situation of this experiment was somewhat special. Therefore, before using the model parameters obtained in this study for other simulations, one should contemplate whether this other situation is comparable to the one presented here. Nevertheless, in this contribution, it worked out well to re-use the parameters from the first experiment to simulate also the second one, where an additional class participated.

8.5 APPENDIX A: EQUATIONS

Equation 8.1 gives the probability for cell (x, y) to be chosen as the destination cell. The details of the influences from the static (p^S) and dynamic (p^D) floor field, as well as from inertia (p^I), the walls (p^W), and other agents (p^P) are explained in Appendix A.

$$p_{xy} = N p_{xy}^S p_{xy}^D p_{xy}^I p_{xy}^W p_{xy}^P$$

Equation 8.2 gives the probability of agent A choosing exit E. $\delta_{AE} = 1$ for the exit that was chosen by agent A at the last round, k_E is a coupling constant that has to be set to fit the simulation to the circumstances of an evacuation.

$$p_E^A = N \frac{1 + \delta_{AE} k_E (A)}{S(A,E)^2}$$

Equation 8.3 gives the partial probability from the static floor field S, that cell (x, y) is chosen as the destination cell. S_{xy}^e contains the information on the distance of cell (x, y) to exit e, k_S is a coupling constant that has to be set to fit the simulation to the circumstances of an evacuation.

$$p_{xy}^S = e^{-k_S S_{xy}^e}$$

Equation 8.4 gives the partial probability from the static floor field D, that cell (x, y) is chosen as the destination cell. (a, b) is the current position of agent A and (x, y) the position of the cell in focus. k_D is a coupling constant that has to be set to fit the simulation to the circumstances of an evacuation.

$$p_{xy}^D = e^{k_D (D_x (x,y)(x-a) + D_y (x,y)(y-b))}$$

Equation 8.5 gives the partial probability, from the inertia, that cell (x, y) is chosen as the destination cell. $(\Delta x_t, \Delta y_t)$ is the velocity vector of the last round, and $(\Delta x_{t+1}, \Delta y_{t+1})$ is the possible velocity vector of this round (the vector for which still has to be decided. v_{last} and v_{next} are the corresponding absolute values (rounded to integers), and k_I is a coupling constant, that has to be set to fit the simulation to the circumstances of an evacuation. A derivation of this equation can be found in (Kretz, 2007).

$$p^I (\Delta x_{t+1}, \Delta y_{t+1}) = \exp \left(-k_I (v_{next} + v_{last}) \sqrt{\frac{1}{2} \left(1 - \frac{\binom{\Delta x_{t+1}}{\Delta y_{t+1}} \binom{\Delta x_t}{\Delta y_t}}{\left| \binom{\Delta x_{t+1}}{\Delta y_{t+1}} \right| \left| \binom{\Delta x_t}{\Delta y_t} \right|} \right)} \right)$$

Equation 8.6 gives the partial probability, from possible nearby walls, that cell (x, y) is chosen as destination cell. W_{xy} is W_0 minus the distance to the wall closest to agent A. If W_{xy} is larger than the cut-off parameter W_0, W_{xy} is set to 0. k_W is a coupling constant that has to be set to fit the simulation to the circumstances of an evacuation.

$$p_{xy}^W = e^{(-k_W W_{xy})}$$

Equation 8.7 gives the partial probability, from possible nearby agents, that cell (x, y) is chosen as the destination cell. N_P is the number of cells within the Moore neighbourhood of (x, y) occupied by other agents. k_P is a coupling constant that has to be set to fit the simulation to the circumstances of an evacuation.

$$p_{xy}^P = e^{-k_P N_P(x,y)}$$

8.6 ACKNOWLEDGMENTS

This work was financed by the Bundesministerium für Bildung und Forschung (BMBF) within the PeSOS project. The authors wish to thank Ms. Dommers, headmaster of the primary school "Am Knappert" in Duisburg-Rahm, and her colleagues for their cooperation. Furthermore, Anna Grünebohm, Frank Königstein, Florian Mazur, and Mareike Quessel are gratefully acknowledged for their support during the evacuation exercise and the evaluation process.

8.7 REFERENCES

Blue, V. J., and Adler, J. L. (2000) "Cellular automata microsimulation of bi-directional pedestrian flows", *TRR, TRB*, 1678:135–141.

Burstedde, C., Klauck, K., Schadschneider, A., and Zittarz, J. (2001) "Simulation of pedestrian dynamics using a 2-dimensional cellular automaton". *Physica A*, 295:507. arXiv:cond-mat/0102397.

Fukui, M., Sugiyama, Y., Schreckenberg, M., and Wolf, D. E. (editors) (2002) *Traffic and Granular Flow '01*, Springer.

Helbing, D., and Molnar, P. (1995) "Social force model for pedestrian dynamics". *Physical Review E*, 51:4282–4286. arXiv:cond-mat/9805244.

Helbing, D., Farkas, I. J., Molnar, P., and Vicsek, T. (2006) *Simulation of Pedestrian Crowds in Normal and Evacuation Situations*. In Schreckenberg and Sharma, eds., pp. 21–58.

ASERI (2008) www. ist–net.de.

Kashiwagi, T. (editor) (1994) *Fire Safety Science – 4th International Symposium Proceedings*, Interscience Communications Ltd, West Yard House, Guildford Grove, London. The International Association for Fire Safety Science.

Kirchner, A., Nishinari, K., and Schadschneider, A. (2003) "Friction effects and clogging in a cellular automaton model for pedestrian dynamics", *Physical Review E*, 67(056122). arXiv:cond-mat/0209383.

Klüpfel, H. (2003) "A cellular automaton model for crowd movement and egress simulation". PhD Thesis, Universität Duisburg-Essen.

Klüpfel, H., and Meyer-König, T. (2002) *PedGo Users' Manual*.

Klüpfel, H., Meyer-König, T., and Schreckenberg, M. (2002) "Comparison of an evacuation exercise in a primary school to simulation results". In (Fukui et al., 2002), pp. 549–554.

Kretz, T. (2007) *Pedestrian Traffic – Simulation and Experiments*. PhD Thesis, Universität Duisburg-Essen.

Kretz, T., and Schreckenberg, M. (2006) *Moore and More and Symmetry*. In (Waldau et al., 2006), pp. 297–308. arXiv:0804.0318.

Kretz, T., Wölki, M., and Schreckenberg, M. (2006) "Characterizing correlations of flow oscillations at bottlenecks", *Journal of Statistical Mechanics: Theory and Experiment*, P02005. arXiv: cond-mat/0601021.

Nishinari, K., Kirchner, A., Namazi, A., and Schadschneider, A. (2004) "Extended floor field CA model for evacuation dynamics", IEICE Transactions on Information and Systems, E87-D:726–732. arXiv: cond-mat/0306262.

Predtetschenski, W. M., and Milinski, A. I. (1971) *Personenströme in* Gebäuden. Berechnungsmethoden für die Projektierung. Verlagsgesellschaft Rudolf Müller, Köln-Braunsfeld (in German, translation from Russian).

PTV – Planung Transport Verkehr AG. (2008) *VISSIM 5.1 User Manual.*

Rogsch, C. (2005) *Vergleichende Untersuchung zur dynamischen Simulation von Personenströmen.* Master's Thesis, Bergische Universität Wuppertal (in German).

Schadschneider, A. (2002) *Cellular Automaton Approach to Pedestrian Dynamics – Theory.* In (Schreckenberg and Sharma, 2002), pp. 75–85. arXiv:cond-mat/0112117.

Schadschneider, A., Klingsch, W., Klüpfel, H., Kretz, T., Rogsch, C., and Seyfried, A (2009) *Evacuation Dynamics: Empirical Results, Modeling and Applications.* In R. A. Meyers, editor, Encyclopedia of Complexity and System Science. Springer, to be published April 2009. arXiv:0802.1620v1.

Schreckenberg, M., and Sharma, S. D. (editors) (2002). *Pedestrian and Evacuation Dynamics*, Springer, Berlin Heidelberg.

Thompson, P. A., and Marchant, E. W. (1994) *Simulex; Developing New Computer Modelling Techniques for Evaluation.* In (Kashiwagi, 1994), pp. 613–624.

Waldau, N., Gattermann, P., Knoflacher, H., and Schreckenberg, M. (editors) (2006) *Pedestrian and Evacuation Dynamics '05.* Springer, Heidelberg.

CHAPTER 9

INCORPORATING PATTERN-MATCHING INTO A DATA-ORIENTED ACTIVITY SIMULATION USING PROBE PERSON SYSTEMS

Eiji Hato, Yasuo Asakuara, Masuo Kashiwadani

This contribution describes the attempt to first transform location data obtained from a mobile communication system into sequences that represent certain behavioral contexts, thereby converting an enormous amount of location data into an object that can be directly operated and analyzed. Subsequently, the behavioral patterns were analyzed in urban spaces. Although the activity pattern sequences that have been considered in conventional studies can be regarded as rough sequences obtained on the basis of questionnaires, the present study is characteristic in that it has aimed at developing an efficient matching method for behavioral patterns intended for sequences with detailed space-time resolutions for the data oriented simulation. Furthermore, the concept of association rule has been incorporated into the data-oriented simulation model, and the memory problems were solved using hash. No discrepancy was found in the simulation results.

9.1 BACKGROUND AND OBJECTIVES OF STUDY

In recent years, information technologies such as RFID (**R**adio **F**requency **ID**entification) tags and mobile phones have been generating dramatic changes in the flow of the quantity and quality of traffic-economy data. Such changes are promoting the reexamination of models that have conventionally been used in the fields of economics and traffic engineering and are generating a practical need for data-oriented methodologies for traffic-economy analysis, such as data mining. Marketing models that evaluate microscopic economic activities of people and control theories for improving traffic congestion depend on the acquisition accuracy and amount of data as well as on computer performances. Therefore, the data revolution by mobile communication, which generates a large amount of accurate data flow, is not unrelated to the paradigm shift in these models and analytical frameworks.

If the conventional data observation approaches based on questionnaires, etc., which tried to capture and model the movement of travelers by focusing on a zone or a cross section, are called Euler-like approaches, the probe person survey, which longitudinally measures the movement of travelers themselves as dots, can be defined as a Lagrange-like approach. If the probe person survey using a mobile communication system enables the observation of Lagrange-like travel-activity data and computer capacity improves, it will become possible to handle a traffic assignment model that explicitly deals with path traffic volume, which has conventionally been dealt with as an intermediate variable. Moreover, the introduction of the cross-validation check with behavioral data on a space-time network for a number of days will change the method for confirming model reproducibility. The possibility of observing day-to-day travel behaviors and associated fluctuations in the state of the network will essentially promote the understanding of traffic phenomena and may lead to the review of set model frames themselves.

Marketing analysis and traffic analysis based on such location data obtained by a mobile communication system can be divided into steps of (1) data measurement, (2) data cleaning, (3) data analysis, and (4) simulation analysis.

The amount of location data collected through data measurement is enormous, and it is difficult to convert it into analyzable data because of the high data-handling load. To solve such a problem, several proposals have been made, such as a basic correction algorithm for location data using PHS (Personal Handy phone System) (Hato and Asakura, 2001) and an efficient method for constructing a location database predicated on OLAP (On-Line Analytical Processing) based on Σ-TREE. Next, the data thus corrected will undergo data analysis and then be used for simulation analysis. Data-oriented methods for behavioral analysis based on large volumes of location data have been proposed, such as cloning simulation, in which data is increased by extrapolation, and spatial data mining. An emphasis of these studies is to develop a method of data analysis that is appropriate and efficient relative to the accuracy and amount of data obtained. It is impossible to evaluate the effect of traffic information in reducing the time required for travel at an accuracy of ±5 minutes by using results of questionnaire responses from subjects on time measurements with decomposition accuracies of ~ ±10–15 minutes. A framework of a model according to observable data accuracy is thus required. Moreover, location data based on a mobile communication system involves an enormous amount of transactions. In addition to conventional data, an efficient method of analysis is therefore of utmost importance.

The present study has attempted to first transform location data obtained from a mobile communication system into sequences that represent certain behavioral contexts, thereby converting an enormous amount of location data into an object that can be directly operated and analyzed. The next step of the study was to analyze the behavioral patterns in urban spaces. Before the analysis of location data, existing studies of the methods of analysis of travel-activity patterns were summarized. This travel-activity pattern analysis has been based on the classification of behavioral contexts. In certain cases, major methods of analysis were utilized for analyzing each attribute value and the characteristics of individuals constituting the cluster for each cluster obtained by classification of behavioral contexts by using cluster analysis. In other cases, the methods were used for measuring the goodness-of-fit when evaluating the degree to which an observed travel-activity pattern was predictable by a model.

In such approaches, the concept of the degree of similarity was used as the classification index for behavioral patterns. In defining the degree of similarity, it is important to consider the notation for the order and intervals of travel-activity patterns as well as consecutive travel-activity patterns and their interdependency relationships. In early studies, Hanson (1982) and Pas (1984) analyzed the associations between the characteristics of a specific individual or household and its travel-activity patterns and demonstrated that typical activity patterns can be explained by the individual's characteristics of location of residence and socio-economic attributes. Recently, Joh et al. (2002) proposed a method able to incorporate information on the sequence of activities and their continuity and that takes into account the interdependency between dimensions, which is a basis for the patterns.

Unlike such existing studies, the present study has converted location data obtained by a mobile communication system into pattern sequences that represent behavioral contexts, identified and operated them, and then developed an algorithm for judging the degree of similarity between patterns. Figure 9.1 displays the data flow in the analysis of travel-activity patterns using location data from a mobile communication system. As can be seen, there are three layers in the simulation analysis flow. The database layer and the data processing layer can be characterized as Probe Person responsive methods. Although the activity pattern sequences that have been considered in conventional studies can be regarded as rough sequences obtained on the basis of questionnaires, the present study is characteristic in that it was aimed at developing an efficient matching method for behavioral patterns intended for sequences with detailed space-time resolutions.

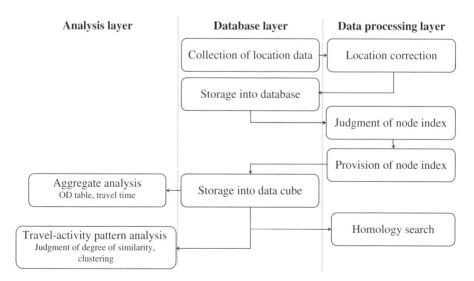

Fig. 9.1 The data flow of travel-activity pattern analyses using location data from a mobile communication system.

9.2   PREPROCESSING OF LOCATION DATA

9.2.1   Data processing

Location data collected using for instance a mobile communication system is defined by Eq. (9.1) as data that shows the location of a traveling object i in an urban four-dimensional space (latitude, longitude, height, and time).

$$\text{Data} = \{i, x, y, z, t\} \qquad (9.1)$$

The direct use of raw data for the analysis of the dot data represented by Eq. (9.1) should decrease the efficiency of data processing, which is associated with data size. Since the amount of location data obtained by using a mobile communication system is enormous, it is not realistic to employ such data directly for analysis, even if improvements in computer performance are taken into account. Certain measures must be taken if real-time processing, such as OLAP (On-Line Analytical Processing), is to be assumed.

9.2.2   Node index and data cube storage

By introducing the concept of node index, the present study attempts to compress data into a data scale with which behavioral contexts can be analyzed. As variables for identifying the behavioral contexts stay ($A_i|i$ = facility type) and travel ($T_j|j$ = travel type) were considered, and node indexes were defined by these variables.

In identifying stay A and travel T, the travel distance per unit time was taken into account. If the distance d between two temporally consecutive observation points is smaller than a previously given threshold D, it is judged that no travel was performed between time t and time $t + 1$. Threshold D is represented by a function (Eq. (9.2)) of the observation time interval Δt between two points and observation location accuracy r.

$$D = f(\Delta t, r) \qquad (9.2)$$

In an x-y-t space, as the one shown in Figure 9.2, threshold D represents the radius of a cylinder drawn around dot data. D can be set using observation error depending on the type of location positioning systems (Hato and Asakura, 2001). When taking into account that the identification of stay/travel is performed by online processing in the order it is written into the database, two types, i.e., the provisional node index and the definitive node index, are considered. Dot data written at temporal point t labels the provisional node index by Eq. (9.2), using its distance d from the dot data written at temporal point $t - 1$. Next, if a new dot data is observed at temporal point $t + 1$, the node index at point t is judged to be T if the distance d between the two points is larger than threshold D; the node index at point t is judged to be A if the distance d between the two points is smaller than threshold D. In addition, the facility type and travel type are judged by using the spatial addresses based on RFID and the results of spectrum analysis by acceleration chips.

Subsequently, the dot data of the traveling object labeled with node indexes was stored in a data cube. At this time, data is held in each node while the node indexes

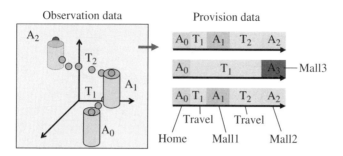

Fig. 9.2 Provision of node index.

to which the object belongs (travel, stay, and facility type) were verified. The nodes were structured by the analysis unit. Data, such as the coordinate area $(x, y, t) \subset L_u$ of the node, velocity sum, time sum V_{Lu}, and the number of objects N_{Lu} in the node, were stored in upper node L_u. The structural storing of the data into a node rendered it possible to obtain the data of all objects in the node through a single access to it. When data is stored in an arbitrary node (area, path, time frame, individual, facility, etc.) to be analyzed, the data is compressed, and it becomes possible to efficiently interpret dot data in a certain behavioral context.

9.3   PATTERN-MATCHING METHOD FOR LOCATION DATA

Using location data provided with node indexes, methods for analyzing similar characteristics that exist in travel-activity patterns when comparing a set of travel-activity sequences that have a certain length are considered. Figure 9.3 shows data obtained by providing node indexes for location data and converting them into travel-activity sequences. The element numbers of the sequences are times, and location flags with which behavioral contexts can be interpreted are substituted into numerical values. As pattern-matching methods for such a set of travel-activity sequences, the multidimensional scale classification method, the dot-matrix method, and the scoring model method were taken into account.

9.3.1   Dot-matrix method

In the dot-matrix method, the sequences to be compared, shown in Figure 9.2, were arranged in both columns and rows, and a matrix taking a value of 1 when corresponding elements matched and a value of 0 when they did not, was generated. This is depicted in Figure 9.4, where the areas that have homology between travel-activity sequences are shaded by hatching. Since uncertainty in decision-making on the time axis cannot be taken into account in these areas, the areas including adjacent k elements (shaded areas) were considered to be homologous. The degree of match M (%), i.e., a scale of homology between two sequences, was taken as the percentage of the number of matching elements in the homologous areas.

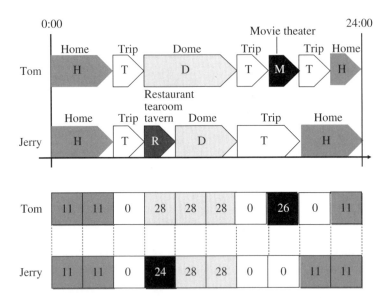

Fig. 9.3 Travel-activity sequences.

Jerry

Tom \ Jerry	11	11	0	24	28	28	0	0	11	11
11	1	1							1	1
11	1	1							1	1
0			1				1	1		
28					1	1				
28					1	1				
28					1	1				
0			1				1	1		
26										
0			1				1	1		
11	1	1								1

Fig. 9.4 Dot-matrix methods.

9.3.2 Scoring model method

The dot-matrix method cannot express the weight of a match between corresponding elements. For example, regarding behaviors occurring over a day, the time passed at a convenience store is shorter than the time passed at home. Therefore, the probability that two persons match by being at a convenience store at a certain time of day is lower than the probability that they match by being at home at a certain time of day. In order to take this weight into account, the present study introduced the concept of the scoring model.

Definitions of notation: The two sequences to be compared are sequences X and Y, and the lengths of the sequences are m and n, respectively. x_i is the ith code of X, and y_j is the jth code of Y. Here, the case of $m = n$ is considered.

Scores based on the scale of log likelihood represented by (there is some association between sequences)/(there is no association between sequences) are assigned to two given sequences. In this case, for the two sequences, the probability that there is some association between them and the probability that there is none are estimated and the ratio between these probabilities is considered.

First, the random model R, where there is no association between sequences, is considered. Assuming that code a is independently observed at a frequency of q_a, the probability that the two given sequences are coincidentally observed is the product of the observation frequencies of the code at each location of the two sequences.

$$P(X,Y \mid R) = \prod_{i=0}^{m} q_{x_i} \prod_{j=0}^{n} q_{y_j} \tag{9.3}$$

Next, the match model M, where there is some association between sequences, is considered. In the match model, if the tth code of X is a and the tth code of Y is b, assume that the tth-code pair of the two sequences is observed at a joint probability p_{ab} (the probability of observing a and b at the same position in the code)

$$P(X,Y \mid M) = \prod_{i=0}^{n} p_{x_i y_i} \tag{9.4}$$

The likelihood ratio of these two equations is defined by the Eq. (9.5) as an odds ratio

$$\frac{P(X,Y \mid M)}{P(X,Y \mid R)} = \frac{\displaystyle\prod_{i=0}^{n} p_{x_i y_i}}{\displaystyle\prod_{i=0}^{n} q_{x_i} \prod_{i=0}^{n} q_{y_i}} = \prod_{i=0}^{n} \frac{p_{x_i y_i}}{q_{x_i} q_{y_i}} \tag{9.5}$$

where an additive scoring system called the log odds ratio S can be derived by taking the logarithm of the odds ratio. Here, the log odds ratio S is defined as the degree of match between two sequences.

$$S = \sum_{i=0}^{n} s(x_i, y_i) \tag{9.6}$$

where

$$s(a,b) = \log\left(\frac{p_{ab}}{q_a q_b}\right) \tag{9.7}$$

is the log likelihood ratio of the code pair probability without association between sequences to the code pair probability with some association between sequences. Here, the log likelihood ratio $s(a, b)$ is defined as the homology score of code a of sequence X and code b of sequence Y.

9.3.3   Multidimensional scale alignment method

The multidimensional scale alignment method is a technique for determining the total value of pattern-matching costs for travel-activity sequences by expressing the costs of matching the sequences in a matrix and searching for the shortest path within it. Figure 9.5 presents an example of such a calculation. In this figure, the same element number of row and column is provided with the stay facility type of the travel-activity sequence in the same time frame. Since the procedures for making the sequences of the facility attributes are identical, four types are considered: "deletion," "insertion," "substitution," and "match." As far as procedure costs are concerned, 1 is taken for either deletion or insertion, 2 for substitution, since substitution is a procedure in which deletion and insertion are simultaneously performed, and 0 represents a match, since a match is a mere confirmation of the fact that the element values of the source pattern and the target pattern are identical.

In the example in Figure 9.5, two travel-activity sequences, i.e. 213 and 991, are compared. In this case, the shortest path cost of matching the two sequences was 4. In contrast, in the case of 311 and 111, the shortest path cost was 2 since the travel-activity sequences could be matched by merely replacing the 3 in the first time frame with 1. Thus, in the multi-dimensional scale alignment method, a shortest path search was performed on a procedure cost matrix, the obtained shortest path cost was defined as a homology score, and the set of sequence patterns that had the smallest score was considered to have the highest degree of match.

Out of the three techniques, the multidimensional scale alignment method is one that only considers sequence orders. If activity choices on the time axis are not taken into account, such a method is suitable as it directly calculates the cost of rescheduling.

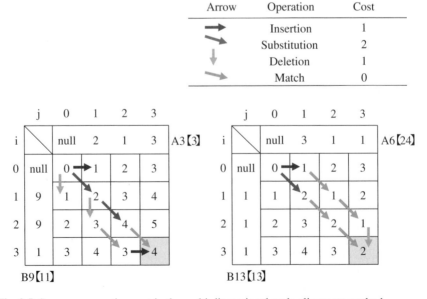

Fig. 9.5 Sequence operation cost in the multi-dimensional scale alignment method.

In contrast, in some cases, a difference in time has an importance in analysis, even if the sequence is the same. When one wishes to analyze the variability pattern of for instance a transport demand that concentrates in a short time, a suitable method is one that looks precisely at time matches between sequence elements, e.g., the dot-matrix method or the scoring model method. The present study has compared and examined the dot-matrix method and the scoring model method while taking into account the high time accuracy of location data from mobile communication systems and the fact that the data consisted of location data at a large-scale event used for a case study.

9.4   CASE STUDY

The characteristics of the proposed methods were examined by using actual data, i.e., the location data of 44 persons between 7:00 and 19:30 on Saturday, November 24, 2001, on which a J. League game was held at the Sapporo Dome (starting time of the game: 14:00). This data was obtained with a mobile communication system. The location data was provided with node indexes and is visually expressed in Figure 9.6, in which the colors represent facilities, and the lengths correspond to activity times. The central green represents the stay of the subjects in the Sapporo Dome. It can be seen that they remained in the Sapporo Dome during the time of the game and that the variance in the time of travel was smaller after the game had ended as opposed to before it had started.

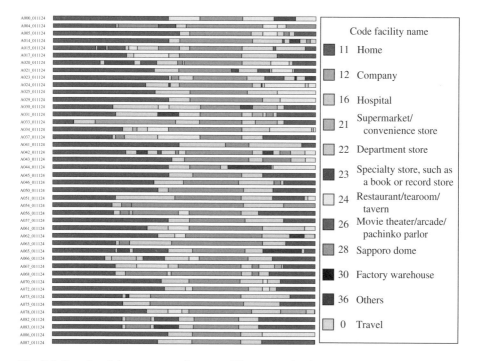

Fig. 9.6 Travel-activity sequences from a mobile communication system.

The dot-matrix method and the scoring model method were applied to the 44 survey samples, and the degrees of match between the different individuals were calculated. Figure 9.7 shows each degree of match according to the calculations. In the dot-matrix method, diagonal components indicate the degree of match of 100%. In contrast, in the scoring model method, even diagonal components demonstrate score differences as a result of the homology score being the odds ratio of the random model to the matching model. In the scoring model method, the odds ratio of travel-activity sequences at an interval of 1 minute for the random model and the matching model was calculated as the homology score. Therefore, if the normal probability of the random model was high, the score became comparatively low.

Next, for each method, the pair with the highest degree of pattern match was extracted, and homology ranks for each pair were obtained by calculating the scores according to the various methods (Fig. 9.8).

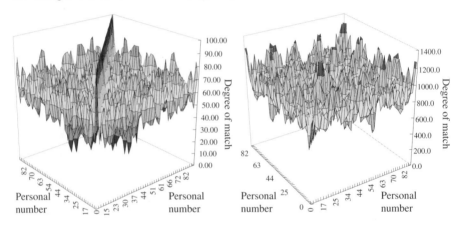

Fig. 9.7 A comparison of the homology scores. Left: Dot-matrix method; Right: Scoring model method.

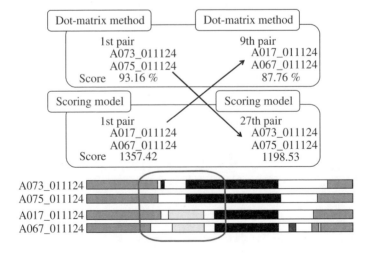

Fig. 9.8 Index comparisons of algorithms.

Although the pair of A073_011124 and A075_011124 displayed the highest homology according to the dot-matrix method, this pair presented the 27th homology score according to the scoring model method, ranking below the pair of A017_011124 and A067_011124 (which had a low score according to the dot-matrix method). It can be said that the pair of A017_011124 and A067_011124 displayed a high homology score according to the scoring model method due to the pair staying at a restaurant/tearoom for about 1 hour before watching the J. League game at the Sapporo Dome, thus showing matches in multiple activity locations. Conversely, it appears that the pair of 073_011124 and A075_011124 demonstrated a high score as a result of the pair taking a simple trip pattern of home → Sapporo Dome → home and the degrees of match in departure and arrival times thus being high.

9.5   SIMULATION STUDY

A simulation model that performs a clustering of time-space base sequences based on a result of homology analysis of these sequences and calculates convergence while revising OD matrix to be correct to link pedestrian flow was proposed. Figure 9.9 shows the data flow in the simulation process consisting of four phases: the first phase concerned data monitoring, the second phase generated processing data with the help of raw data, and in the fourth phase, a data analysis, such as a simulation or an OD estimate based on the result, handled data with the third phase.

The present study proposed the tour simulation based on the association rule using the observation data with high resolution. Ordinary approaches using questionnaires employ the model that aggregates trips by zones and represent excursion tours as decomposition trips. On the other hand, the probe person system renders it possible to represent the travelers' microscopic excursion trips due to an identification

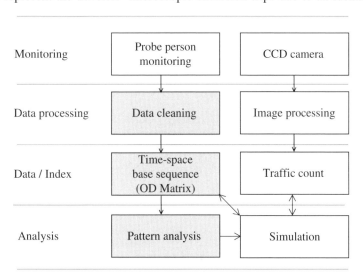

Fig. 9.9 The flow of data in a data-oriented simulation.

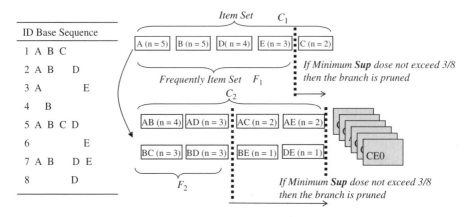

Fig. 9.10 Enumeration algorithms of excursion patterns.

of detailed traveler trajectories. It is important that the number of conditions in the probability of choosing facilities becomes massive, which can be achieved by increasing the number of observed facilities that respondents visit. The combination number in the conditional probability of choosing stops thus soon exceeds giga byte size. These facts lead to a multitude of problems such as the increment of computing memory in the simulation calculation, and it becomes difficult to obtain a solution. The conditional probabilities for emerging these massive patterns need to be calculated. The threshold dose of enumerating the conditional pattern is considered; particularly the association rule mining using hash with minimum support has been applied to this study. At first, base sequences (i.e., the respondents' patterns of stops) were made, and the support for the set of items was calculated. If the minimum support of the item did not exceed a certain value, the branch was pruned (Eq. (9.8)).

$$F = \left\{ X \subset I \,|\, \sup(X) > \min \sup \right\} \qquad (9.8)$$

By running over these computation processes, there is usually no need for the enumeration of massive patterns. Next, the conditional probability $P(j|i)$ was defined using Eq. (9.9) and the parameter was estimated.

$$\ln P(j|i) = \left(\sum_l \beta_l \cdot \delta_l \right) \ln T_j - \gamma \ln d_{ij} + \left(\sum k'_{ij} \cdot \delta_{ij} \right) + \left(\sum k''_{ij} \cdot \delta_{ij} \right) + k \qquad (9.9)$$

$$T_{ij} = T_i \cdot P_{j|i} = \frac{T_i^{\alpha} \cdot T_j^{\beta}}{d_{ij}^{\gamma}} + k' + k'' \qquad (9.10)$$

where:

$P(j|i)$	:	Confidence
T_i	:	Number of customers at the shop i
T_j	:	Number of customers at the shop j
d_{ij}	:	Distance between the shop j and the shop j
k', k'	:	Dummy parameter of behavioral context

k :   Constant

α_k, β_l :   The parameter of facilities' attributes (k,l)

$\delta_k, \delta_l, \delta_{ij}$:   Dummy Variables

Figure 9.11 shows a pedestrian network that was used for the case study. A numerical computation was carried out for an excursion pattern of pedestrians in the Matsuyama area. In the case studies, the homology distance of the base sequence of an excursion pattern was calculated according to the method described in Section 9.4 and simulated using the results of cluster analysis based on the homology distance. Then followed an evaluation of the simulation model using observed pedestrian numbers at several places on the roadside. Table 9.1 shows the estimation results, and the parameter of distances indicates significant values that would be logically consistent. Moreover, the parameter of behavioral context was incorporated into Eq. (9.10), and the results indicated a significant value. The probability of an emerging excursion among variant facilities depended on a certain combination of facilities. Figure 9.12 shows the results of an OD estimation using the simulation model, and it was seen that the difference between the simulation value and the actual survey value was larger at the link with a few pedestrian counts as opposed to at other links. In this simulation model, its elasticity was not entirely low.

9.6   CONCLUSION

Methodologies for directly operating location data collected at an interval of 1 minute by using a mobile communication system and analyzing the matching of behavioral patterns were compared and examined. An empirical analysis revealed that the dot-matrix method and the scoring model method varied with regard to characteristics and that the dot-matrix method was suitable for analyses that concerned the times of occurrence of detailed trips, whereas the scoring model method was suitable for analyzing

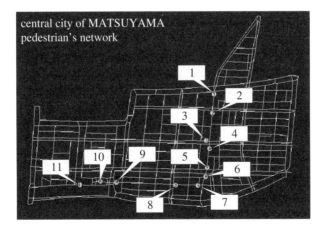

Fig. 9.11 The network and the observation point for the OD estimation.

Table 9.1 Estimation results.

Parameter				t-Statistics
Constant	k		−6.758	(−18.49)
Destination	β_1	Commercial complex	0.196	(3.89)
	β_2	Clothing store	0.652	(5.88)
	β_3	Café shop/restaurant	0.194	(1.69)
	β_4	Book store	0.231	(3.24)
	β_5	Drag store	0.228	(2.52)
	β_6	Recreational facilities	0.995	(5.75)
	β_7	Bank	0.145	(1.04)
	β_8	Park	0.818	(1.25)
	β_9	Others	0.030	(0.22)
	γ	Intergroup distance	0.154	(2.53)
Facility context	k'_{21}	Clothing store–Commercial complex	1.018	(3.55)
	k'_{31}	Restaurant–Commercial complex	0.993	(3.39)
	k'_{52}	Recreational facilities–Restaurant	1.693	(4.17)
Area context	k''_{12}	A1–A2	−0.984	(−4.73)
	k''_{13}	A1–A3	−1.111	(−4.46)
	k''_{43}	A4–A3	−1.584	(−8.91)
		Sample number	539	
		R^2	0.953	

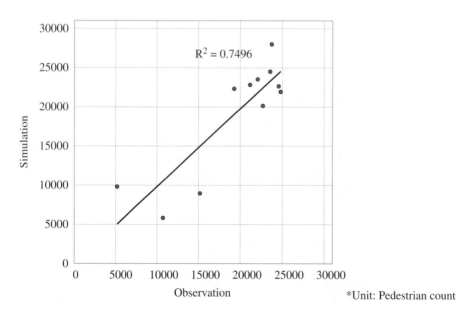

Fig. 9.12 Simulation results.

diverse travel-activity patterns, such as urban marketing analyses. The proposed simulation model could coincidentally analyze massive travel-activity patterns after carry-
ing out the data cleaning for the dot data. Moreover, when obtaining high-resolution
data, the behavior context representing the concatenation of multiple behavioral units
became complex. Association rule methods were proposed as a means of dealing with
this problem, and such an approach showed a good degree of calibration as well as a
decent reproducibility. Furthermore, in cases where the pedestrian count was adopted
as a fit index, good results could be obtained with the data-oriented simulation model.

9.7 REFERENCES

Hanson, S. (1982) "The determinants of daily travel-activity patterns: relative location and socio-demographic
    factors", *Urban Geography*, 3(3):179–202.
Hato, E., and Asakura, Y. (2001) *New Approach for Collection of Activity Diary Using Mobile
    Communication Systems*. CD-ROM, TRB Annual Meeting, Washington, D.C.
Joh, C.-H., Arentze, T., Hofman, F., and Timmermans, H. (2002) "Activity pattern similarity: a multidimen-
    sional sequence alignment method", *Transportation Research part B*, 36(5):385–403.
Pas, E. I. (1984) "The effect of selected sociodemographic characteristics on daily travel-activity behavior",
    Environment and Planning A, 16:571–581.

CHAPTER 10

SIMULATION OF URBAN RAIL OPERATIONS: MODELS AND CALIBRATION METHODOLOGY

Haris N. Koutsopoulos, Zhigao Wang

SimMETRO is a microscopic, dynamic, stochastic simulator of urban rail operations (METRO), specifically designed for service performance analysis, and the evaluation of operations and strategies for a real-time control of subway systems. SimMETRO employs a detailed representation of the network, rolling stock, signal control, demand, schedule and dispatching. In particular, the various sources of stochasticity in operations are explicitly captured. This chapter presents approaches for the calibration of model parameters and input (such as dynamic arrival and alighting rates), and a case study illustrates the applicability of the model and the proposed calibration methodology. After calibration, the RMSE of block run times (relative to actual times as reported by the train detection system) was reduced by 60% as compared to the default values. The calibrated demand was in good agreement with recent counts of arrival rates at the stations.

10.1 INTRODUCTION

There are a number of rail simulation models (see e.g., Bendfeldt et al., 2000; Goodman et al., 1998; Nash and Huerlimann, 2004; RAILNET II, 2005) that mainly focus on planning and train performance analysis for general rail systems. The application of rail simulation models requires the calibration of these models in order for them to accurately replicate observed conditions in the system. However, the literature on calibration and validation of rail simulation models is very limited. Existing approaches for rail model calibration are not very advanced, and most methods are ad hoc and use simple statistics or performance measures to compare the simulator output to field observations while adjusting the model parameters, by trial and error, until the simulated measurements are close to the observed ones (Tromp, 2004; Venglar et al., 1995; White, 2005). Such approaches may work satisfactorily in cases where there is little or no uncertainty incorporated in the simulation, or when most of the parameters or input data are of good quality and reliability. However, when the number of parameters increases and various sources of randomness are present, existing methods are unable to calibrate the models in a systematic and consistent manner.

The calibration of road traffic simulation models, on the other hand, has experienced considerable progress in recent years (see e.g., Balakrishna, 2006; Balakrishna et al., 2005, 2006, 2007; Ben-Akiva et al., 2004). These methods are quite general and can be used for the calibration of a number of model parameters and inputs such as dynamic Origin-Destination flows, capacities, etc. Hence, there is a potential in employing similar approaches for the calibration of rail simulation models.

The present chapter focuses on the simulation of urban heavy rail (METRO) operations and its objective is twofold:

- to introduce a new simulation model, SimMETRO, for mass transit subway systems, particularly its features in representing various sources of randomness that impact on operations; and

- to discuss general methods for the calibration of the important parameters and necessary input of these models using commonly available data, such as track (block) detection data.

The remainder of the chapter is organized as follows. Section 10.2 introduces SimMETRO and presents the main design characteristics supporting the desired functionality. Section 10.3 details the calibration methodology, and Section 10.4 illustrates the application of SimMETRO as well as the calibration methodology through a case study of the operations of the Red Line in Boston, MA. Finally, Section 10.5 concludes the chapter.

10.2   THE SIMULATION MODEL

SimMETRO is a microscopic, dynamic, stochastic simulation model, specifically designed for service performance analysis, as well as the evaluation of operations and strategies for real-time control of subway systems. The intended applications of SimMETRO include:

- operations planning;

- system performance analysis;

- development and evaluation of real-time operations control strategies;

- capacity analysis;

- training.

In order for SimMETRO to provide the required functionality, the design demands include the following capabilities:

- a dynamic representation of the requirement and its impact on train operations (e.g. dwell times at stations);

- a representation of the main sources of uncertainty including, demand, incidents (i.e. train malfunctions, station emergencies);

- a delay propagation through trip chaining and train schedules;

- individual operator performance characteristics;

- train performance and characteristics;

- control system uncertainties;

- a flexibility in preparing schedule inputs.

The structure of the simulation model is illustrated in Figure 10.1.

The *network* representation captures merging points and junctions, in addition to all geometric features that impact train performance (e.g. grade and curvature). The network also includes the characteristics of the stations, such as their length and platform configuration.

The *train control system* represents the corresponding block design and associated speed code. The control system regulates train movements and maintains safe separation. The track is divided into circuit blocks. When a train occupies a track

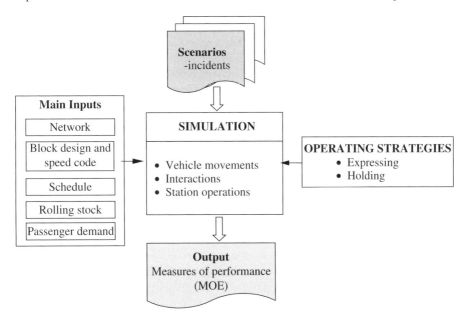

Fig. 10.1 The architecture of the simulation model.

block, the following train is typically prohibited from entering the preceding track block, thus maintaining a minimum of one clear block between trains. A track boundary is a signal box that sends a signal upstream through the rails and receives the signal by the following signal box. Each block has a different signal frequency than its neighbors, and there is thus no confusion regarding the signal origin (Fig. 10.2). As the train proceeds and the block occupancy changes, the speed code changes accordingly. The occupied block sends speed indications for the trains occupying upstream blocks, and these indications govern the speed at which the trains are allowed to move downstream.

Various types of train control can be simulated, including fixed block control. At any time, a fixed block signaling system knows only whether a block is occupied or clear. There is no information about where the train is positioned within the occupied block.

Blocks are adequate for maintaining train separation on a single track. However, at merging or diverging locations and in areas with multiple rights of way, an additional control is required to prevent trains from conducting conflicting movements. The inter-locking of blocks is implemented for this purpose.

The *schedule* can be specified by a number of options, depending on the application. For example, it may constitute an input in the form of departure headways and their time-dependent distributions, in the form of scheduled headways (including the composition of trains and the trips they perform), or in the form of actual headways (to simulate the operations of a specific day). In addition, the schedule includes information concerning not only headways and/or departure times for the trips, but also trip chains (train schedule) so that delays can be accurately propagated as trains perform their sequence of trips. The structure of a detailed schedule is given in Figure 10.3.

The *rolling stock* is represented with all the operating characteristics that impact train performance, such as the seating capacity, the total capacity, the door configuration, the car composition, and acceleration/ deceleration profiles. Figure 10.4 illustrates an example of train acceleration as a function of the speed of the train.

Depending on the type of control used, the behavior of the train operator may have an impact on the actual acceleration/deceleration used as well as on the resultant speed. This impact can be captured in SimMETRO, through operator-specific parameters.

Passenger demand can be modeled at various levels of detail, depending on the available data. Under the simplest option (when information concerning passenger arrival rates or origin-destination flows is unavailable), the impact of demand is captured by dwell times at stations. These dwell times are station-specific and random.

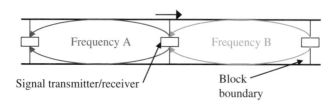

Fig. 10.2 A schematic of the signal equipment layout.

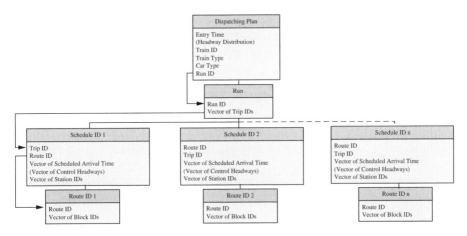

Fig. 10.3 The structure of a schedule.

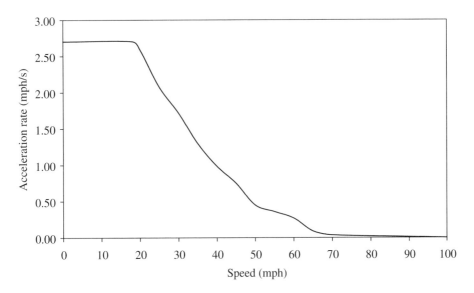

Fig. 10.4 An example of a train acceleration profile.

If information on passenger arrival/alighting rates is available, a detailed dwell time model is employed, which estimates the dwell time as a function of the number of boarding and alighting passengers. Section 10.4, Eq. (10.5) presents an example of dwell time models used in the simulation. The dwell time model is flexible and can accommodate any functional form. In this case, the number of arrivals is determined according to a stochastic process (e.g. Poisson). The default arrival model assumes that passenger arrivals follow a non-homogeneous Poisson process:

$$P\big[B(t, t + \Delta t) = k\big] = \frac{(\lambda(t)\Delta t)^k e^{-\lambda(t)\Delta t}}{k!} \qquad (10.1)$$

where, $P[B(t, t + \Delta t) = k]$ is the probability that k passengers arrive within the interval Δt, and $\lambda(t)$ is the time-dependent arrival rate. Depending on available data, other models that possibly offer a better reflection of actual arrival patterns may also be implemented so as to represent more realistic processes and enhanced estimates of passenger waiting time, crowding, and impact on the level of service; in short, a more accurate reflection of system performance.

Various *operating strategies* for real-time control of operations aiming at service restoration (in the case of a major disruption) and schedule maintenance can be evaluated and refined. Examples include expressing, holding trains at selected stations, or short-turning trains.

SimMETRO is designed to test and evaluate the operating performance of a system under different schedules, control system configurations and concepts, block designs, rolling stock characteristics, etc. In order to evaluate the system, various *Scenarios* can be generated representing possible operating conditions. For instance, a scenario may correspond to a variety of levels of demand. Furthermore, a scenario may include incidents that can be either location- or vehicle-specific. Location-specific incidents are used to represent situations of specific durations, such as a medical emergency at a station or an equipment malfunction at a certain block. Train-specific incidents are used to represent a disabled train. In such cases, the train will first slow down, then stop for a predefined duration, after which the activities can be specified for that train. This includes the unloading of passengers at the first downstream station it arrives at, expressing the subject to a lower speed limit (than the speed code), and leaving the network.

In order to evaluate a design or operating policy, SimMETRO outputs a rich set of *performance measures* (MOE) both related to system and the level of service, such as travel times, headway distribution, passenger waiting times, train passenger loads, number of passengers unable to board the first train, etc. These MOEs are system-, route segment-, stop-, vehicle- or, passenger-based.

The associated MOEs can be used to assess the performance of the system both from a productivity point of view and from the perspective of the passenger level of service (LOS), including traditional measures such as waiting time, as well as other important measures such as reliability.

Figure 10.5 illustrates the graphical user interface (GUI) of SimMETRO used to animate train movements and facilitate the assessment of the system's performance.

Through its detailed nature, SimMETRO captures the uncertainty in urban rail operations by explicitly representing the various sources of stochasticity in the system. Such sources include levels of demand, deviations from schedule, the propagation of delays through the system, operator variability as well as its impact on train performance, and train interactions.

10.3   CALIBRATION METHODOLOGY

The application of a model such as SimMETRO requires the calibration of model parameters and inputs for the system under investigation. Important parameters

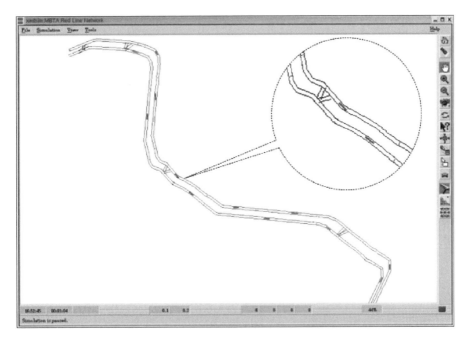

Fig. 10.5 The SimMETRO graphical user interface.

include those of the dwell time model as well as the train acceleration/deceleration profiles (if unknown). Time-dependent (dynamic) arrival/alighting rates at stations are critical input variables that may be unavailable; especially for older systems. Data for calibration can be obtained from a variety of sources:

- train control and data acquisition system (OCS) data, in the form of activation and deactivation times for each block, and hence station occupancy times: these occupancy times are representative of the dwell time, and are functions of the (unknown) arrival and alighting rates;

- prior studies of system performance (e.g. dwell time investigations) that provide a-priori estimates of the corresponding model parameters;

- prior passenger surveys and demand counts at various stations.

Depending on the type of data and its source, the above information may provide direct or indirect measurements of the input variables and parameters of interest at various levels of accuracy. Prior surveys provide direct measurements of demand levels at various stations. Dwell times at stations give indirect measurements of demand, since they are dependent on the boarding and alighting passengers. The train control and data acquisition system (OCS) permits a direct measure of the activation and deactivation time of each block in the system due to the passage of a train. Hence, this data can provide valuable information concerning headways at various locations in the system, as well as travel times, and dwell times at stations.

A general methodology for the estimation of urban rail simulation parameters and inputs from track activation data and other sources is proposed, based on a formulation of the issue as an optimization problem. The objective was to minimize both errors (i.e. differences between simulated and observed values) and deviations of model parameters and input variables from a-priori values.

$$\text{Min } Z = \sum_{i=1}^{m} w_{1i}(Y_i^{obs} - Y_i^{sim})^2 + \sum_{j=1}^{n} w_{2j}(P_j^a - P_j)^2 + \sum_{k=1}^{l} w_{3k}(I_k^a - I_k)^2$$

$$st. \tag{10.2}$$

$$Y^{sim} = S(P^a, I^a)$$

Here,

P, P^a	:	vectors of calibrated and a-priori system parameters, e.g. dwell time model parameters, with members P_j, P_j^a;
I, I^a	:	vectors of calibrated and a-priori system inputs, e.g. time dependent arrival/alighting rates, with members I_k, I_k^a;
Y^{sim}, Y^{obs}	:	vector of simulated and observed measurements, e.g. dwell time at stations, with members Y_i^{sim}, Y_i^{obs};
w_{ij}	:	weights capturing the importance and accuracy of each component;
$S(P^a, I^a)$	:	a simulation model that maps the inputs to the corresponding measurements;
m	:	the number of observation points (stations) multiplied by the time periods;
n	:	the number of parameters;
l	:	the number of stations multiplied by the time periods.

The values of the weights w_i reflect the importance of each term in the objective function. For instance, measurements with high errors will have small weights and hence minor contributions. The weights are determined through the application of generalized least squares (GLS) (see Balakrishna et al., 2005; Pindyck and Rubinfeld, 1997).

The calibration model is a multivariate, stochastic optimization problem, and the problem is difficult to solve since it is simulation-based and hence without a closed form (analytical) objective function (function $S(.)$ mapping the inputs to the measurements is, in this case, the simulation model, SimMETRO). The SPSA algorithm has demonstrated very good performances in the application reported in this chapter (Spall, 1998).

10.4   CASE STUDY

The applicability of SimMETRO and the developed calibration methodology was demonstrated through a case study involving the operations of the Red Line of the Massachusetts Bay Transportation Authority (MBTA) in Boston, USA (Fig. 10.6).

The Red Line configuration is an older system operating under automatic train operations (ATO) control. Operators have a certain amount of flexibility in their choice of acceleration/deceleration as well as speed (under the received speed command).

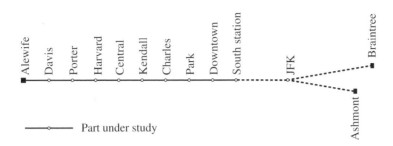

Fig. 10.6 The Red Line configuration.

The line operates with two main train types, series 1800 and series 1700, which vary with regard to acceleration/deceleration capabilities. During peak hours, trains typically consist of 6 cars with 18 or 24 doors (depending on the train type) and have a capacity of about 960 passengers (including standees).

The case study focuses on the southbound, afternoon peak period (4 PM–6 PM), and the available data for the case study includes OCS data from October of 2004; older count data on arrival and alighting rates at the stations, by time of day, collected manually at the stations in 1997 (but not separated by branch); and dwell time model parameters calibrated from actual data collected at select stations in 2000. Since the OCS data was from October 2004, the model was calibrated for that period.

Based on the block activation/deactivation times, departure times at the terminal station, headways at various stations, travel times, and dwell times were calculated for each trip. The dwell times from the OCS data represented an approximation of the true dwell time as they included the time for the train to enter the station (Dixon, 2006).

During the afternoon peak, the line was very congested and operated at scheduled headways of mostly 4 minutes. Moreover, the reliability of the line was rather low. Figure 10.7 compares the actual departure headways at the southbound terminal (as recorded by the OCS system) to the scheduled headways for the 4 PM–6 PM time period, and illustrates the broad divergence in actual headways.

Based on the OCS data and the scheduled headways, the distribution of departure headways was developed separately for three sub-periods: 4 PM–4:40 PM, 4:40 PM–5:15 PM and 5:15 PM–6 PM. Figure 10.8 illustrates the empirical distribution of departure headways for two of these three time periods.

SimMETRO was calibrated with the above inputs. The focus of the calibration was on the station-specific parameters of the dwell time model and the time-dependent arrival/alighting rates of passengers at each station by branch. The arrival/alighting rates were estimated for 30-minute intervals, and the choice of parameters and input variables to calibrate reflected the relative importance of these factors, the sensitivity of the results to those values, and the fact that reliable estimates of the parameters were unavailable from other sources. Hence, the parameters were very critical for the accuracy of the model and presented a great opportunity for improvement.

Following the general formulation of the calibration model presented in Section 10.3, the objective function used was the minimization of the (square) error between the simulated and actual dwell time at stops and the (square) difference between arrival/alighting rates and dwell time model parameters and their a-priori values.

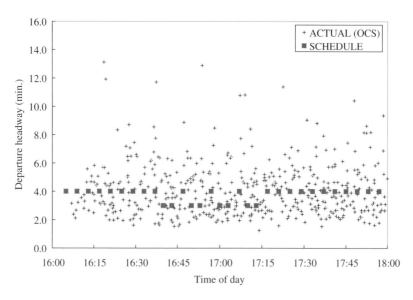

Fig. 10.7 The scheduled vs. actual departure headway distribution.

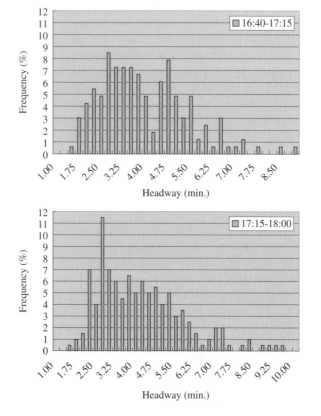

Fig. 10.8 An empirical distribution of dispatching headways.

The available demand data that served as the a-priori values was collected in 1997. Clearly, due to changes in economy, demography, and land use since that time, the 1997 counts could only provide limited information concerning the 2004 arrival/alighting rates. Furthermore, the 1997 data represented total counts and were not sorted by branch, while the input to the simulation required time-dependent arrival/alighting rates by branch.

The dwell time model used in this case study was based on a model developed in an earlier investigation (Puong, 2000):

$$D = C + \beta \cdot B + \alpha \cdot A + \gamma \cdot TS^3 \cdot B \tag{10.3}$$

where,

D	:	the dwell time
C	:	a constant with a calibrated value of 12.22 seconds,
A	:	the number of alighting passengers per door,
B	:	the number of boarding passengers per door,
TS	:	the number of through standees per door,
α, β, γ	:	parameters with calibrated values of 2.27 (s/pax), 1.82 (s/pax) and 0.00064, respectively.

The model was calibrated with a relatively small sample collected at two of the Red Line stations (South Station and Kendall/MIT). Hence, the calibrated parameters should capture the dwell time at those stations fairly well. However, in some cases, the remaining stations had quite different characteristics that impacted the dwell time, such as the location of passenger accesses, the size and number of platforms, the level of platform congestion and the passenger distribution on the platform. Since SimMETRO was able to accommodate various dwell time models for different stations, station-specific parameters were calibrated using these numbers as a-priori values.

In order to evaluate the effectiveness of the calibration, the estimated passenger arrival rates obtained from the calibration were compared to the actual arrival rates in 2004 at select stations for which such data was available. It should be noted that only the 1997 demand data rates were used for calibration (as a-priori values), and that the 2004 data was used exclusively for validation purposes after the calibration.

Figure 10.9 compares the calibrated arrival rates at the four stations for which data from 2004 was available. For each station, the reported arrival rates corresponded to 30-minute intervals for the 4 PM–6 PM peak period. The graph illustrates the a-priori values (1997), the actual values (2004), as well as the calibrated values for the period from 4 PM–6 PM, in 30-minute intervals, and the results indicate that the calibration provided relatively good estimates for the actual arrival rates, especially for high demand stations. When the 1997 data was used, the corresponding RMSE (over all stations and time periods) was 247 pax/hr. The RMSE of the calibrated demand was 107 pax/hr, corresponding to a 57% reduction.

Table 10.1 summarizes the calibration error for each station and time period (percent error of calibrated and actual arrival rates). In general, predictions were better for stations with high levels of demand. This was expected since the actual time needed for boarding/alighting passengers at these stations dominated the dwell time, while data from stations with low demand levels presented rather high levels of noise.

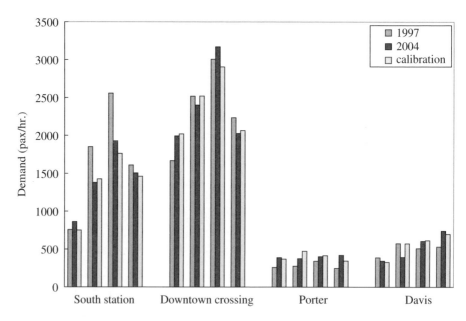

Fig. 10.9 A comparison of arrival rates.

Table 10.1 The calibration error in the arrival rates (%).

Time period	4:00–4:30	4:30–5:00	5:00–5:30	5:30–6:00
South station	−13%	+3	−9	−3
Downtown crossing	+1	+5	−8	+2
Porter	−5	+27	+3	−18
Davis	−5	+46	+1	−6

Table 10.2 summarizes the parameter values from the calibrated dwell time model for the various stations and compares them against the original model (independently calibrated with data collected specifically for that purpose at two of the stations). In general, the boarding and alighting times per passenger were slightly adjusted by the calibration, while the constants changed substantially for certain stations due to the station characteristics differing from those of the stations where the original model was calibrated. For example, Park station has two platforms that are used for passenger loading and unloading. At this station the operator has to attend to doors on both sides and hence spend more time on door operations. The extra time associated with the operations at Park was captured by the higher value of the constant. In other cases, the constant could absorb some of the time that a train would stand at a station due to congestion, which otherwise might not be captured by the simulation model (e.g. Charles station). Note that, for the calibration, the parameter of the congestion term in the dwell time model was fixed to its original value.

Table 10.2 Calibrated values for dwell model parameters.

Parameter	C (s)	α (s/pax)	β (s/pax)
A-priori values (Puong model)	12.22	2.27	1.82
Davis	15.23	2.40	1.89
Porter	13.61	2.09	1.84
Harvard	18.12	2.04	1.83
Central	16.71	2.29	1.77
Kendall	13.91	2.34	1.83
Charles	20.04	2.41	1.88
Park	23.83	2.52	1.89
Downtown	14.14	2.37	1.87
South station	17.40	2.34	1.83

In addition, and in order to further evaluate the performance of the calibration approach and the validity of the simulator, a number of goodness of fit statistics were calculated including the Root Mean Square Error (RMSE), and Theil's U inequality (for more details see Pindyck and Rubinfeld, 1997; Toledo and Koutsopoulos, 2004):

$$\text{RMSE} = \sqrt{\frac{1}{N}\sum_{n=1}^{N}(Y_n^{\text{sim}} - Y_n^{\text{obs}})^2} \qquad (10.4)$$

$$U = \frac{\sqrt{\dfrac{1}{N}\sum_{n=1}^{N}(Y_n^{\text{sim}} - Y_n^{\text{obs}})^2}}{\sqrt{\dfrac{1}{N}\sum_{n=1}^{N}(Y_n^{\text{sim}})^2} + \sqrt{\dfrac{1}{N}\sum_{n=1}^{N}(Y_n^{\text{obs}})^2}} \qquad (10.5)$$

Y_n^{sim} and Y_n^{obs} are here the simulated and observed measurements, respectively, (time- and location-specific). U values of 0 indicate a perfect fit, while $U = 1$ corresponds to a bad performance. U can be decomposed into three proportions: U^M (bias), U^S (variance), and U^C (covariance), and for a good model, U^M and U^S should be as small as possible, while U^C should be close to 1. Values of U^M that are higher than 0.10 indicate a high degree of bias.

The calibration was found to significantly improve the performance of the model. Table 10.3 shows the goodness of fit results with respect to block run times (i.e. the time when a block is activated to the time it is deactivated). As can be seen, the RMSE was reduced by 60%. The values of Theil's U and its proportions also indicated an overall improved ability to replicate existing conditions. Most importantly, the U^M value showed a significant reduction in bias.

Figure 10.10 displays the scatter plot of average block run times before and after calibration. Prior to calibration, the runtimes were biased, in particular for blocks

Table 10.3 Calibration statistics on block run times.

Statistic	Before	After
RMSE (s)	4.455	1.767
U (Theil's statistic)	0.081	0.030
U^M (bias)	0.204	0.036
U^S (variance)	0.302	0.031
U^C (covariance)	0.494	0.933

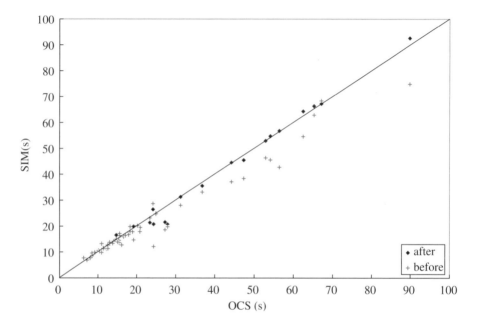

Fig. 10.10 A comparison of the block run time.

with longer run times (blocks with stations where dwell times contribute to the majority of the run times), or blocks where congestion occurred. After the calibration, a significant improvement was gained suggesting that the errors in the dwell times and congestion were corrected.

While the replication of block run times is a meaningful indication of the simulation model's ability to represent operations in detail, other aspects are equally important. For example, a correct distribution of headways over time and across the line is critical as it results from interactions among trains and uncertainties in the operations. The headway distribution is a comprehensive indication of whether trains are moving and spaced correctly. Furthermore, the headway distribution affects the calculation of the level of service measures, such as waiting times.

Figure 10.11 shows the arrival headway distribution before (top) and after (bottom) calibration at Park Street station, where one of the highest demands occurs. The calibrated simulation model replicated the observed headways well.

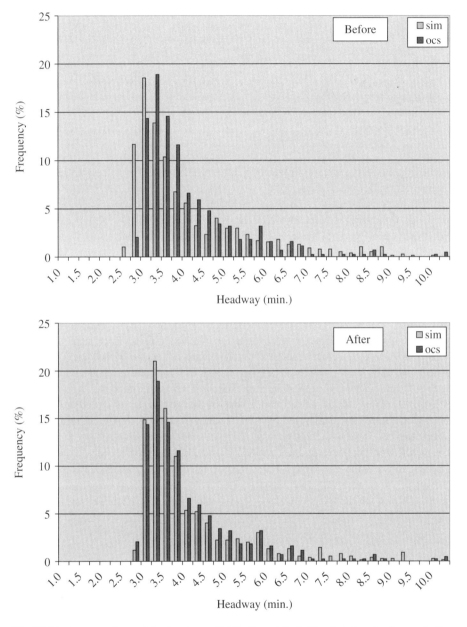

Fig. 10.11 A comparison of the headway distribution at Park Street station (before and after calibration)

Finally, the calibrated model was validated against block run time data from a day that was not used in the calibration process. During this day, trains were dispatched from the terminal at the actual departure times and sequence as reported by the OCS data for the 4 PM–6 PM time period. Table 10.4 summarizes the results. For

Table 10.4 Validation statistics for block run times (single day).

Statistic	Before	After
RMSE (s)	3.55	2.75
U (Theil's statistic)	0.057	0.042
U_M (bias part)	0.097	0.001
U_S (variance part)	0.145	0.042
U_C (covariance part)	0.76	0.957

reference, Table 10.4 also includes the results from the model using the default values (before calibration).

The statistics after calibration corresponded well with observed values and there was a clear improvement when comparing to the statistics before calibration, especially with respect to the bias, which was substantially reduced.

10.5   CONCLUSIONS

Urban heavy rail simulation models are useful tools for the analysis of rapid rail transit operations and for evaluating control strategies. SimMETRO is a fast transit simulation model, specifically designed to simulate such systems at the operational level, capturing important details, such as the uncertainty in operations. The calibration of such models is a very important activity before their actual use, and can be facilitated by the availability of extensive data on block activation/deactivation times. A calibration methodology that takes advantage of the availability of block occupancy data to estimate important model parameters and inputs, such as passenger arrival rates at stations, has been presented. The results from a case study indicated that this calibration methodology employed the available information effectively in order to estimate the input variables and parameters (arrival/alighting rates, and dwell model coefficients). Moreover, it was able to improve the accuracy of the simulation results. The calibrated SimMETRO model was thus capable of replicating observed conditions very accurately.

10.6   ACKNOWLEDGMENTS

The present research was partially supported by NSF, project CMS-0339108, and a grant from MBTA. The authors would like to thank MBTA for the data used in this study.

10.7   REFERENCES

Balakrishna, R. (2006) *Off-line Calibration of Dynamic Traffic Assignment Models*, PhD Thesis, MA, Institute of Technology.

Balakrishna, R., Ben-Akiva, M., and Koutsopoulos, H. N. (2006) "Time-dependent origin-destination estimation without assignment matrices", in *Proceedings* (*cd-rom*) *of the 2nd International Symposium of Transport Simulation* (*ISTS*), Lausanne Switzerland, September.

Balakrishna, R., Ben-Akiva, M., and Koutsopoulos, H. N. (2007) "Off-line calibration of dynamic traffic assignment: simultaneous demand-supply estimation", *Transportation Research Record,* 2003:50–58.

Balakrishna R., Koutsopoulos, H. N., and Ben-Akiva, M. (2005) "Calibration and Validation of Dynamic Traffic Assignment Systems," Transportation and Traffic Theory (ISTTT): "*Transportation and Traffic Theory*", H. Mahmassani editor, pp. 407–426, Pergamon Press.

Ben-Akiva, M., Darda, D., Jha, M., Koutsopoulos, H. N., and Toledo, T. (2004) "Calibration of Microscopic Traffic Simulation Models with Aggregate Data", *Transportation Research Record*, 1876:10–19.

Bendfeldt, J. P., Mohr, U., and Muller, L. (2000) *RailSys, a System to Plan Future Railway Needs*, Computers in Railways VII, WIT Press.

Dixon, M. (2006) *Analysis of a Subway Operations Control Database: The MBTA Operations Control System*, MS Thesis, Northeastern University, Boston, MA.

Goodman, C. J., Siu, L. K., and Ho, T. K. (1998) *A Review of Simulation Models for Railway Systems*, International Conference on Developments in Mass Transit Systems.

Nash, A., and Huerlimann, D. (2004) *Railroad Simulation Using OpenTrack*, Computers in Railways IX, WIT Press.

Pindyck, R. S., and Rubinfeld, D. L. (1997) *Econometric Models and Economic Forecasts*, 4th edition. Irwin McGraw-Hill, Boston MA.

Puong, A. (2000) *Dwell Time Model and Analysis for the MBTA Red Line*, MIT Research Memo.

RAILNET II (2005) http://www.fasta.ch/railnet_II/railnet_II_english.htm. Accessed May.

Spall, J. C. (1998) "An overview of the simultaneous perturbation method of efficient optimization", *Johns Hopkins APL Technical Digest*, 19:482–492.

Toledo, T., and Koutsopoulos, H. N. (2004) "Statistical validation of traffic simulation models", *Transportation Research Record*, 1876:142–150.

Tromp, J. P. M. (2004) "Validation of a Train Simulation Model with Train Detection Data", Computers in Railways IX, J. Allan, C. A. Brebbia, R. J. Hill, G. Sciutto and S. Sone editors, WIT Press.

Venglar, S. P., Fambro, D. B. and Bauer, T. (1995) "Validation of simulation software for modeling light rail transit", *Transportation Research Record* 1494:161–166.

White, T. (2005) "Alternatives for railroad traffic simulation analysis", *Transportation Research Record* 1916:34–41.

PART IV

Simulation Application

CHAPTER 11

TRAFFIC SIMULATION FOR AN EXPRESSWAY TOLL PLAZA BASED ON MASSIVE VEHICLE TRACKING DATA

Ryota Horiguchi, Takahiro Shitama, Hirokazu Akahane, Jian Xing

The present chapter describes the modeling concept of traffic flow at an expressway toll plaza. It is important that traffic conditions in a toll plaza be evaluated in terms of not only efficiency but also safety. As there was less knowledge of the driving behavior of Electric Toll Collection (ETC) vehicles in a toll plaza, the first step was to carry out a precise survey of vehicle trajectories in the plaza prior to the modeling. This chapter first introduces the outline of massive vehicle tracking with five cameras, after which the modeling concept of the traffic simulation for a toll plaza is explained based on the tracking data. The model was implemented as an extension module of AVENUE, a microscopic traffic simulation model, and was validated. The final section describes the case studies through this simulation model as well as the implications of the study.

11.1 INTRODUCTION

Nowadays, Electric Toll Collection (ETC) systems have become rapidly popularized in Japan. Road authorities expect that more than 80 percent of running vehicles will be equipped with on-board ETC units within a couple of years. To accommodate such a high flow of ETC vehicles, the road authorities have to revise the design of toll plazas in order to keep the flow of vehicles smooth and safe by increasing the number of exclusive ETC lanes, modifying the geometry or enlargening the area of the plaza.

For the sake of aiding road authorities, a traffic simulation model taking into account the interaction between traffic conditions and the design of a toll plaza needs to be developed. Traffic conditions should be evaluated in terms of not only LOS but also safety. As there is less knowledge of the driving behavior of ETC vehicles in toll plazas, the first step was the precise survey of vehicle movements in the plaza prior to the modeling.

The subsequent sections introduce an outline of massive vehicle tracking with five video cameras, after which the modeling concept of the traffic simulation for a toll plaza is explained based on the tracking data. The model was implemented as an extension module of AVENUE (Horiguchi et al., 1996), a microscopic traffic simulation model, and was validated. The final section describes case studies in which this simulation model was used as well as the implications of the study.

11.2 MASSIVE TRACKING OF VEHICLE TRAJECTORIES

11.2.1 Video survey and tracking

In order to quantify the vehicle movements in toll plazas, traffic surveys with precise vehicle tracking techniques based on synchronized multiple video cameras (Akahane and Hatakenaka, 2004) was conducted at the Narashino Toll Plaza of the Higashi-Kanto Expressway.

The size of the toll plaza is approximately 500 meters in width and it comprises 11 lanes at the toll gate section including four exclusive ETC lanes (of which one was closed). As it is difficult to cover a whole plaza of this size with a single video camera, five cameras were installed at the top of the two light towers beside the toll gate, and each camera covered part of the plaza, as shown in Figures 11.1 and 11.2. The

Fig. 11.1 Pictures taken by the five video cameras.

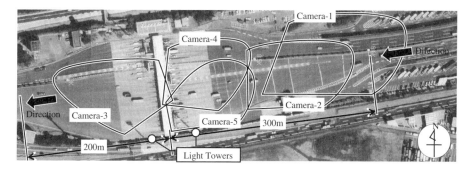

Fig. 11.2 The areas covered by the five video cameras.

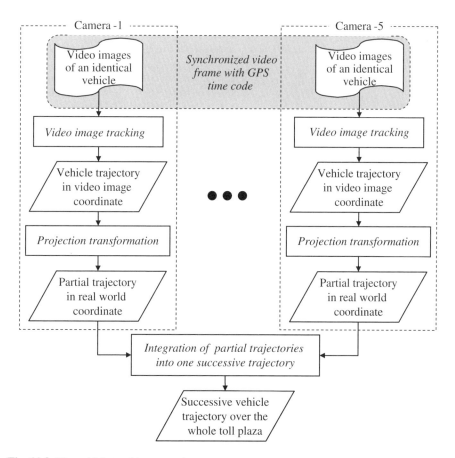

Fig. 11.3 The vehicle tracking procedure.

picture frames of the cameras were synchronized by importing the time code from a GPS satellite signal.

Figure 11.3 illustrates the procedure for estimating a successive vehicle trajectory in the toll plaza as developed by (Akahane and Hatakenaka, 2004). For each video

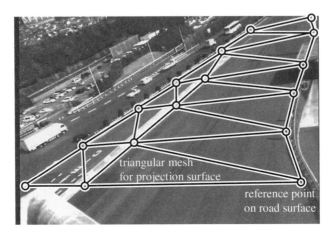

Fig. 11.4 The reference points and triangular meshes on the road surface.

camera, the position of an identical vehicle was tracked to obtain the partial trajectory on the video image coordinate system. Subsequently, the trajectory was projected onto a real world coordinate system with three dimensions. These projected partial trajectories were integrated in one successive trajectory by using Kalman smoother to minimize the location errors, originating from the inaccuracy of the video image processing, the unknown height of the tracking point on the vehicle, etc.

Since the method required the measuring of the location of the reference points on the road surface in both video image coordinates and real world coordinates, certain points at the end of the white lines of the road surface were chosen and their real world locations measured by using a laser surveillance instrument. As shown in Figure 11.4, these reference points formed triangular meshes that were used to identify the projection formula from the video image to reality. Among these meshes, the nearest one to the vehicle position at each moment was used for the projection transformation, as shown in Figure 11.5.

Every 0.1 seconds, the Kalman smoother performed a simultaneous estimation of not only the location but also the kinematic status of the vehicle, such as speed, acceleration, etc. Although certain parts of the plaza were not covered by the cameras, e.g., the toll gate section under the roof, the Kalman smoother was able to estimate the vehicle trajectories according to the vehicle kinematics.

11.2.2   Estimated trajectories

The survey was executed between 10:50 and 12:30 on February 20, 2004, at which time the traffic condition was relatively light. The objective was to capture the lane choice behavior of a vehicle without it being affected by other nearby vehicles. Throughout the 100-minute survey, the trajectories of 1349 vehicles including 344 ETC vehicles were extracted. Figure 11.6 illustrates the vehicle trajectories by an overlay of aerial photographs of the toll plaza. As can be seen, the vehicles arrived at the left lane at the upstream end of the plaza. There were two exclusive ETC gates at the center and two others on the left-hand side (another one to the left was closed). Only the ETC

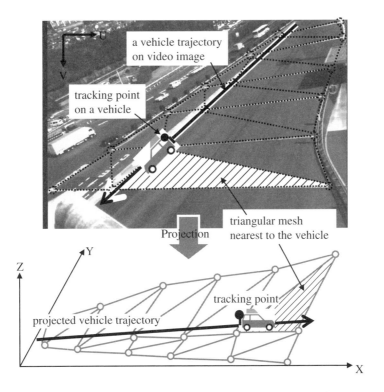

Fig. 11.5 The projection of the vehicle trajectory onto 3-D real world coordinates.

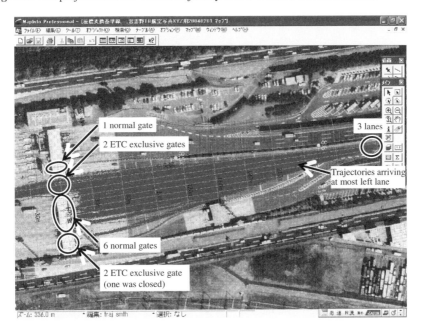

Fig. 11.6 Estimated vehicle trajectories as mapped on an aerial photograph.

vehicles passing through the two ETC exclusive gates at the center were tracked to the downstream of the toll plaza.

11.2.3   Speed and acceleration

Figures 11.7–11.10 illustrate the speed and the acceleration changes on the exclusive ETC lanes as well as on the normal lanes. The gray dots represent the speed

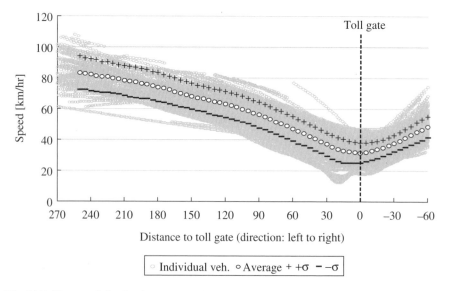

Fig. 11.7 The speed distribution on an exclusive ETC lane.

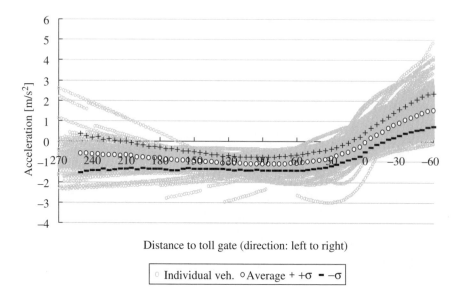

Fig. 11.8 The acceleration distribution on an exclusive ETC lane.

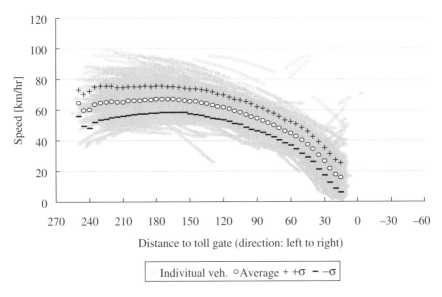

Fig. 11.9 The speed distribution on a normal lane.

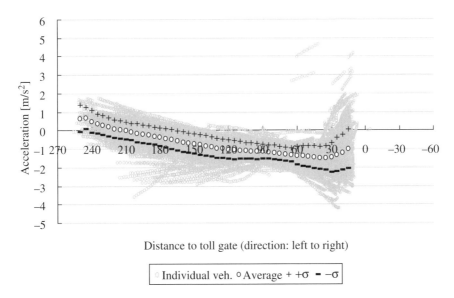

Fig. 11.10 The acceleration distribution on a normal lane.

or the acceleration of an individual vehicle at each position as a function of the distance from the toll gate section. On top of the individual dots are plotted average values and standard deviations (±σ) for every 5 meters. There are some incomplete trajectories for which the speeds or accelerations seemed irregular. These are

Table 11.1 A comparison of the estimated and measured speeds at the ETC toll gate.

	Average	Standard deviation
Estimated speed	32.0 km/hr	7.0 km/hr
Measured speed	34.0 km/hr	7.0 km/hr

mostly due to the imperfect width of the image tracking and failure to be integrated into one complete trajectory by the Kalman smoother. However, since the number of such irregular trajectories was small, they were ignored in the further analysis.

As to the accuracy of the estimated positions, speeds, and accelerations, there was very little reference data available, except for the measured speed at the ETC toll gate. Table 11.1 shows a comparison of the average values and the standard deviation of the passing speed at the toll gate. Although the toll gate section was covered by a roof and thus not filmed by the cameras, the speed distribution of the estimated trajectories agreed well with the measured speed.

With the values from Figures 11.7–11.10, it is possible to understand the average behavior of the approaching vehicles.

- An ETC vehicle arrives at the upper end of the toll plaza at around 80 km/h and gradually slows down with a deceleration of about 1 m/s² up to approx. 60 meters before the gate.

- A normal vehicle arrives at the upper end of the plaza at around 70 km/h, maintains the speed up to 150 meters before the gate, and then gradually slows down.

- The passing speed at the ETC gate was approximately 36 km/h. Bounded by the gate, a vehicle changes deceleration to acceleration, which is more intense than deceleration.

Since the behavior of the vehicles was captured under light traffic conditions, it can be regarded to represent the desired speed and acceleration of dirvers on exclusive ETC lanes. This data will thus be used in the modeling stage.

11.2.4   Headway and capacity

Figure 11.11 shows a histogram of the headway distribution at the ETC toll gate as measured by a vehicle sensor. As can be seen, the headways were smaller than the most frequent values in car-following situations: a curve with a peak situated in the left part of the histogram could represent the headway distribution at high levels of traffic. From the figure, the gamma distribution of which the mean value was 4.5 seemed to correspond well to this left part, and could thus be regarded as the optimum headway to provide a capacity of 800 veh/hr at the ETC toll gate.

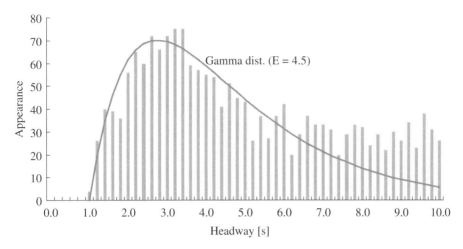

Fig. 11.11 The headway distribution at the ETC toll gate (#10) of Narashino Toll Plaza. Total = 3855 veh/day.

11.3 MODELING OF TRAFFIC FLOW IN TOLL PLAZA

11.3.1 Flow modeling

There are dozens of microscopic simulation models to employ various car-following models which consider drivers' characteristics, such as response delay, desired speed, target headway, etc. However, certain essential parameters seem to be hard to directly obtain from traffic survey data. There are sometimes difficulties in finding clear relationships between these model parameters and the road capacity obtained by the simulation model. Furthermore, the driver characteristics in the simulation model may be less affected by the location on the road, while, in reality, they vary as shown in the previous section.

In order to overcome such difficulties and to fully utilize the result of the video survey, the following basic modeling strategies were employed.

- The car-following behavior was modeled as a first-order status equation, i.e. the headway spacing and speed (S-V) relationship of which the average can be derived from macroscopic observations.

- The average S-V relationship of the car-following behavior was determined by the position in the toll plaza.

- The individual behavior was provided by randomly scaling the average S-V relationship at each position according to the normal distribution of which the variance could be obtained from macroscopic observation.

In this study, a simple S-V relationship, i.e., Greenshield's Formula, was assumed. The formula can be parameterized with the average free flow speed and

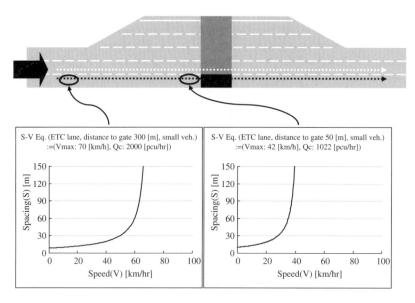

Fig. 11.12 The Spacing-Speed (S-V) relationship at varying positions in the toll plaza.

the capacity at each position in the plaza, which could be obtained through the video survey (Fig. 11.12).

With a car-following behavior based on the S-V relationship, the speed of each vehicle could be calculated with the following simultaneous equations. With this modeling, a consistency was expected between the observed and simulated capacities, which accumulated individual vehicle movements (Yoshii and Kuwahara, 1995).

$$
\begin{cases}
S_i(t+1) = f(V_i(t)) \\
S_i(t) + V_{i-1}(t)dt = S_i(t+1) + V_i(t)dt
\end{cases}
\tag{11.1}
$$

where

f : S-V relationship (macroscopic flow characteristics),
$S_i(t)$: headway distance of vehicle i,
$V_i(t)$: speed of vehicle i,
dt : unit time interval.

According to the survey result, the capacity and the average free flow speed at each location in the ETC and in the normal lanes were determined as shown in Figures 11.13 and 11.14, respectively.

11.3.2 Lane choice behavior in the toll plaza

Inside the toll plaza, a driver can dynamically choose the lane towards the target toll gate, according to the vehicle position, lane operation status (e.g. exclusive ETC, closure), length of queues at the gates, etc. The estimated vehicle trajectories are helpful to identify the lane choice behavior of the drivers.

In order to clarify the following explanation, the lane arrangement in the upstream section of the Narashino Toll Plaza is illustrated in Figure 11.15. The main

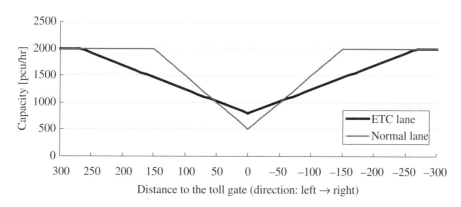

Fig. 11.13 The capacity as a function of the distance to the toll gate at the Narashino toll plaza.

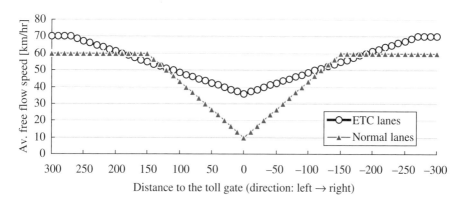

Fig. 11.14 The average free flow speed as a function of the distance to the toll gate at the Narashino toll plaza.

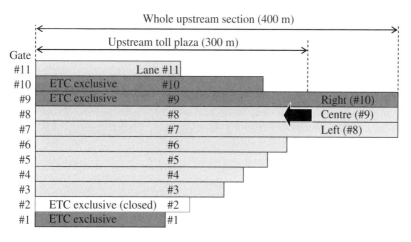

Fig. 11.15 The lane arrangement in the upstream end of the toll plaza.

expressway comprising 3 lanes led vehicles to the upstream end of the toll plaza, 300 meters before the toll gates. There were four exclusive ETC lanes, located in pairs at each end of the plaza. Each gate and lane was numbered from 1 to 11 from left to right as shown in the figure, and as can be seen, gate #9 was in the continuation of the right lane (passing lane) of the main expressway.

Figure 11.16 shows the toll gate choice probabilities of ETC vehicles at different positions in the toll plaza. Each column in the graph displays the probability of a vehicle in a certain position, i.e. 270 m, 210 m or 120 m, choosing a certain toll gate section. For instance, as demonstrated in the top graph in Figure 11.16, more than

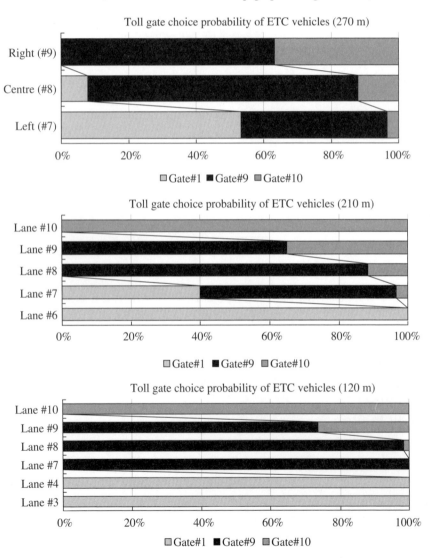

Fig. 11.16 The choice probabilities for the exclusive ETC toll gates at varying positions within the toll plaza.

60% of the ETC vehicles in a the position 270 m from the toll gate chose gate #9 and the rest of them chose gate #10.

When a vehicle was 270 m from the toll gate, the gate choice was rather flexible. However, this flexibility decreased as the vehicle closed in on the gate. For example, the probability vehicles choosing gate #9 when incoming from lane #7 increased from 43% at 270 m, to 57% at 210 m, and finally to 100% at 120 m. Drivers seemed to gradually narrow their choice for the 'target gates'.

Gate #1, which was the farthest to the left, attracted vehicles incoming on lane #6 and #7 to a larger extent than gate #9 which was closer to these lanes. For example, at the 210 m position, 100% of vehicles on lane #6 chose gate #1, despite the fact that it was necessary to change 5 lanes. The drivers who decided to go through gate #1 seemed to do so at an early stage of their approach.

Figure 11.17 is a similar graph for non-ETC vehicles. As for the ETC vehicles, the flexibility of the gate choice gradually decreased as the vehicles came closer to the gates. The gate for which the choice probability was the largest was located slightly to the left of the incoming lane. For example, at the 270-m position, gate #6 had the largest probability of being chosen and this gate was centered in the choice set of the vehicles incoming from the center lane (#8). Such shifts could be the result of the shape of the toll plaza bulging to the left side, thus enabling drivers to find room to approach the gates in this space. Furthermore, the shifting decreased when a vehicle got closer to the gate. At the 120-m position, most of the vehicles chose the gate just in front of them or one lane to the right/left.

From these figures, certain implications should be considered in the modeling phase.

(1) With regard to the ETC gates being split into two pairs, a driver seemed to first determine to go to either the left- or right-hand side of the toll gates when arriving at the upstream end of the toll plaza.

(2) A driver has a wide focus on the choice of toll gates when he/she is far from the gate, and this focus narrows as he/she approaches the gates.

(3) The center of the choice set of the toll gates is shifted to the left of the incoming lane. The shift distance is larger at a position farther from the gates as opposed to near the gates.

11.3.3   Basic structure of lane choice model

Based on the implications above, the choice models comprised three steps, as shown in Figure 11.18. A vehicle arrives from one of the incoming lanes to the upstream end of the toll plaza according to the given choice probability for the arriving lanes. The proportion of each vehicle type, small/large and ETC/non-ETC vehicles, on each lane should also be given.

At the upstream end of the toll plaza, the vehicle determines the selectable lanes to go through. If these selectable lanes are split on both the left and right side and comprise more than two lanes, the 'split side choice model' is applied to delimit

Toll gate choice probability of non-ETC vehicles (270 m)

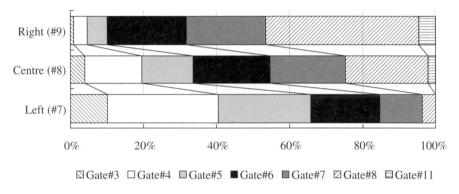

⊠ Gate#3   □ Gate#4   □ Gate#5   ■ Gate#6   ■ Gate#7   ☑ Gate#8   ⊟ Gate#11

Toll gate choice probability of non-ETC vehicles (210 m)

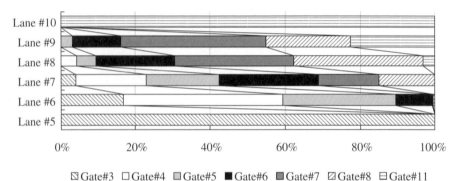

⊠ Gate#3   □ Gate#4   □ Gate#5   ■ Gate#6   ■ Gate#7   ☑ Gate#8   ⊟ Gate#11

Toll gate choice probability of non-ETC vehicles (120 m)

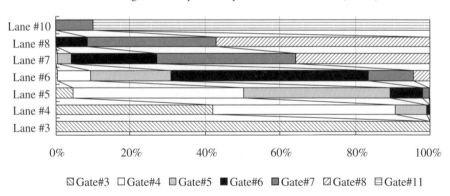

⊠ Gate#3   □ Gate#4   □ Gate#5   ■ Gate#6   ■ Gate#7   ☑ Gate#8   ⊟ Gate#11

Fig. 11.17 The choice probabilities for the normal toll gates at varying positions within the toll plaza.

the selectable lanes on the selected side. Once the vehicle enters the toll plaza, it selects the target gate on one of the selectable lanes according to the '*target gate choice model*' of which parameters may vary depending on the vehicle's position.

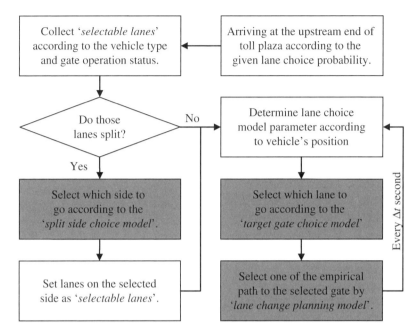

Fig. 11.18 A flow chart of the lane choice behavior of a vehicle.

Subsequently, the vehicle determines one of the possible paths, which consists of the empirical vehicle trajectories estimated through the video survey, to lead it towards the target gate according to the '*lane change planning model*'. Finally, two selection models are repeatedly applied as the vehicle drives in the toll plaza.

11.3.4 Split-side choice model and its calibration

The split-side choice model is described with Eq. (11.2). The choice probability for the lanes on the left side is calculated by a sigmoidal function, of which the exponential term in the denominator consists of the 'lateral offset' from the vehicle's position in the arrival lane to the center position of the left/right side lanes. Figure 11.19 illustrates the variables to be used in Eq. (11.2). Here, the lateral offset is measured in terms of the number of lanes. Moreover, the offset towards the right side constitutes a positive value. In this figure, for the vehicles arriving on the left lane (#7), the value of the lateral offset to the left-side lanes (δ_L) is −6, and the corresponding value to the right (δ_R) is 2.5.

$$P_L = \frac{1}{1+e^{-\theta(\delta_L w_1 - \delta_R w_2)}} \tag{11.2}$$

where

$\quad P_L$: the choice probability for the lanes on the left side.

$\quad \delta_L, \delta_R$: the lateral offset from the vehicle's arrival lane to the center position
                   on the left/right side lanes.

$\quad \theta, w_1, w_2$: parameters to be calibrated.

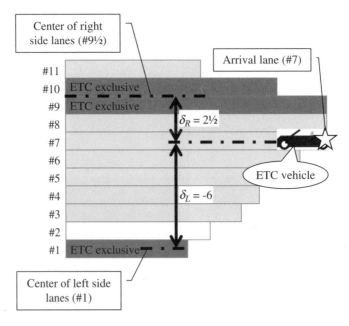

Fig. 11.19 The variables used in the split-side choice model.

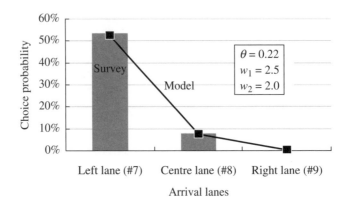

Fig. 11.20 The calibrated split-side choice model for the Narashino toll plaza.

The parameters w_1 and w_2 are weighting factors for establishing balance, and θ refers to the sensitivity. Figure 11.20 depicts the choice probabilities for choosing the left side depending on the arrival lane. The results were obtained from the video survey and compared to the calibrated model with a numerical search for w_1, w_2 and θ. The model was found to fit to the observed choice probability well.

11.3.5   Target gate choice model and its calibration

Once a vehicle enters the toll plaza, it selects a 'target gate' among the selectable lanes, according to the following sigmoidal function (Eq. 11.3).

$$P_x(k) = \frac{1}{1 + e^{-\theta_x(k-\delta_x)}} \qquad (11.3)$$

where

k	:	relative lane position (lateral offset) from the current lane of the vehicle to the target gate. A positive value means an offset to the right.
x	:	normalized longitudinal position in the toll plaza. The position "$x = 0$" is the toll gate section and "$x = 1$" is the upstream end of the toll plaza.
$P_x(k)$	:	cumulative choice probability of lane k from the most left lane to the right.
δ_x	:	a position-dependent parameter to be calibrated (≤ 0).
θ_x	:	a position-dependent parameter to be calibrated (> 0).

As illustrated in Figure 11.21, the parameter δ_x refers to the 'choice center' of the selectable lanes, where the cumulative choice probability from the left side should be 0.5. If δ_x is negative, the choice center is shifted to the left from the current lane position of the vehicle and the vehicle tends to choose the left side to a larger extent. Based on the implications in the lane choice analysis, δ_x is a negative

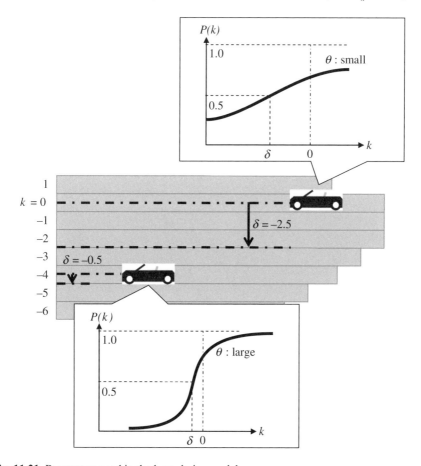

Fig. 11.21 Parameters used in the lane choice model.

value for the Narashino Toll Plaza and its absolute value should become larger for larger x.

The parameter θ_x changes the steepness of the sigmoidal curve. If θ_x is small, the slope of the curve is gentle. This means that the choice probability distribution for each gate is widely spread and that the vehicle – according to such a probability will – demonstrate flexibility when choosing a target gate. On the other hand, a large θ_x value will lead to a loss of this flexibility in the choice behavior. From the implications in the lane choice analysis, θ becomes smaller for larger x.

The parameters for the target gate choice models for ETC and non-ETC vehicles are calibrated by a numerical search and identified according to Eq. (11.4) and (11.5). Figure 11.22 compares the observed choice probabilities at each arrival lane with the estimated probability obtained by the calibrated model for non-ETC vehicles at the 270-m position where "$x = 0.9$".

$$\begin{cases} \theta_x = 4.27e^{-1.14x} \\ \delta_x = -0.56x \end{cases} \qquad \text{(for ETC vehicles)} \qquad (11.4)$$

$$\begin{cases} \theta_x = 4.00e^{-1.50x} \\ \delta_x = -2.68x \end{cases} \qquad \text{(for non-ETC vehicles)} \qquad (11.5)$$

where

x : relative distance to the gate in the upper section of the toll plaza.

In the simulation model, it is necessary to calculate the choice probability of each gate from the cumulative choice probability curve according to Eq. (11.3).

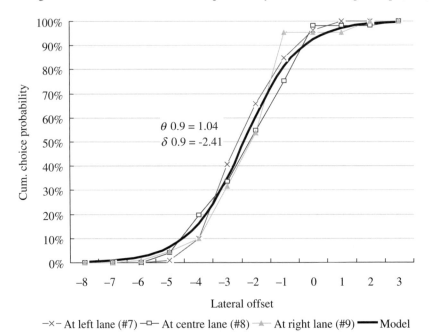

Fig. 11.22 The calibrated target gate choice model for non-ETC vehicles at 270 m.

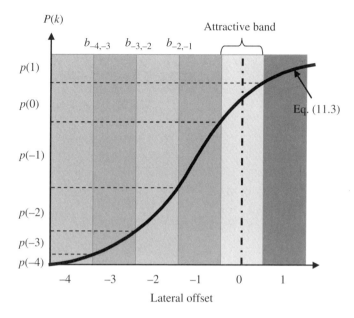

Fig. 11.23 The attractive band for partitioning the cumulative choice probability curve.

Here, we introduce an 'attractive band' to delimit the boundary of the coverage of each lane over the cumulative choice probability curve given by Eq. (11.3). Figure 11.23 illustrates the idea of the attractive band, partitioning the lateral offset axis into the section bounded by $[b_{k-1,k}, b_{k,k+1}]$ for each lane k. The choice probability of the gate at position k can now be calculated by Eq. (11.6).

$$p(k) = \begin{cases} P(b_{k,k+1}), & (k = k_{min}) \\ P(b_{k,k+1}) - P(b_{k-1,k}), & (k_{min} < k < k_{max}) \\ 1 - P(b_{k-1,k}), & (k = k_{max}) \end{cases} \qquad (11.6)$$

where

$b_{k,k+1}$: boundary of the attractive band between k and $k + 1$.
k_{min}, k_{max} : the lateral offset of the most left/right lane.

However, the target gate choice model described so far is ineffective when there are queues in front of the toll gates. Since drivers may prefer the gate for which the queue is the shortest, it is necessary to incorporate the queue length effects into the model.

In order to express such queue length effects, the boundary of each attractive band is calculated by Eq. (11.7) so as to change the width of the attractive band in inverse proportions. According to Eq. (11.7), as shown in Figure 11.24, when all of the queue lengths are either zero or equal to each other, the lateral offset axis is divided with the same band width. However, when the queue lengths are unequal, the attractive band for the gate with the relatively shorter queue will be wider, while the band will be narrower for the gate with the longer queue. This mechanism can work to equalize the queue lengths of each gate, since the narrower attractive band with a long queue should decrease the choice probability for this gate.

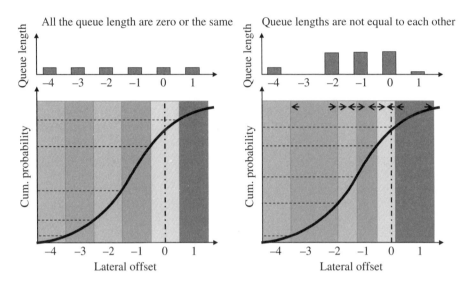

Fig. 11.24 Queue length effects when changing the width of the attractive band.

$$b_{k,k+1} = \begin{cases} k - 0.5, & (k = k_{\min}) \\ b_{k-1,k} + \dfrac{\max(l_k, L_{\min})^{-1}}{\displaystyle\sum_{j=k_{\min}}^{k_{\max}} \max(l_j, L_{\min})^{-1}}, & (k > k_{\min}) \end{cases} \qquad (11.7)$$

where

l_k : queue length on lane k.

$L_{\min}$: insensitive factor for queue length (to avoid zero division).

11.3.6 Lane change planning model

Once a vehicle has determined its target toll gate, it needs to plan a lane changing path to reach this toll gate. However, it seems to be difficult to identify a decision-making model for the drivers based only on the observed vehicle trajectories. In this study, the conventional behavior modeling approach has not been chosen, but rather a data-oriented approach that uses observed trajectories as the choice set of the path plan. This is based on the idea that the drivers' behavior with regard to decision making might be indirectly represented in the real data set.

In the simulation model, all observed trajectories, referred to here as 'paths', are 'embedded' in the toll plaza. At any position in the plaza, a vehicle can pick up one of the paths that passes through that position and leads the vehicle to its target gate, as illustrated in Figure 11.25. The path is randomly selected every 3 seconds at the vehicle's current position. As the selection is an independent phenomenon, two or more vehicles may simultaneously choose the same path.

At the position where the path crosses over the lane-marking line, a vehicle tries to change its lane by following the path. If there is another vehicle driving in the next

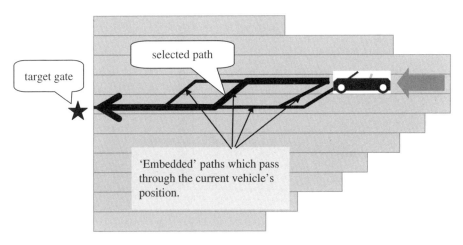

Fig. 11.25 A path selection from 'embedded' trajectories.

lane, the vehicle with the farthest position from the toll gate generally decelerates and gives way. In this sense, the path can be regarded as a 'lane changing plan'.

11.4   VALIDATION AND CASE STUDIES

The simulation model explained in the previous chapter was implemented in an existing traffic simulation software, AVENUE (Horiguchi et al., 1996), as its extension module. Here, the implemented model was validated in terms of lane choice behavior, speed on each lane, and counts of 'near-miss' opportunities.

11.4.1   Reproducibility of lane choice behavior

Figure 11.26 compares the results from the simulation (sim) and observation (obs) with regard to lane choice probabilities for ETC vehicles. The simulation result fit that of the observation well. The probability of the vehicles arriving at the left lane differed from the observation, but this error may have arisen from the limited number of such vehicles.

Figure 11.27 displays similar results for non-ETC vehicles. The choice probabilities for the center gates, i.e., #4 to #7, agreed well with the simulation results but for the gates on either sides, i.e., #3, #8 and #11, the simulation result did not fit those of the observation. Although the parameters δ and θ in the target gate choice model were calibrated as shown in Figure 11.22 in Section 11.3.5, the model curve differed slightly from the observed curves on both sides. This was believed to cause the errors found in Figure 11.27.

Figure 11.28 illustrates the vehicle counts every 10 meters on each lane in the upstream toll plaza. The unit of the numbers in the grayscale cells was [veh/hr], and a darkening of the colors corresponded to an increased magnitude of traffic volume. As shown, the traffic began to diffuse at a distance of 230 m from the gates.

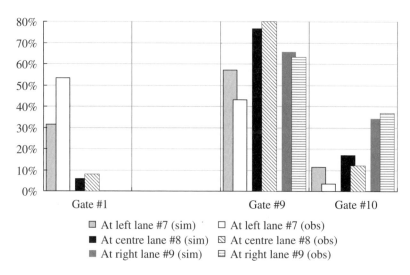

Fig. 11.26 A comparison of the gate choice probabilities for ETC vehicles at 270 m upstream.

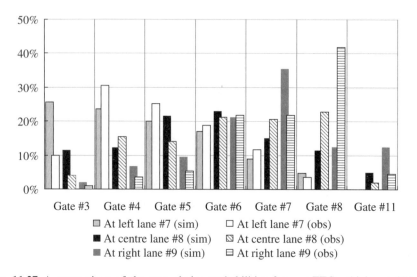

Fig. 11.27 A comparison of the gate choice probabilities for non-ETC vehicles at 270 m upstream.

Gate	0m	10m	20m	30m	40m	50m	60m	70m	80m	90m	100m	110m	120m	130m	140m	150m	160m	170m	180m	190m	200m	210m	220m	230m	240m	250m	260m	270m	280m	290m	300m
#11	36	36	36	36	36	34	34	34	14	14	6	6	6	5	5																
#10	52	52	52	52	43	32	32	30	40	40	38	38	39	32	32	19	19	16	16	16	16	14	14	5	5	0	0				
#9	147	147	147	147	156	165	165	167	122	109	109	109	101	101	98	98	118	118	105	105	115	115	161	161	193	193	192	192	189	189	
#8	57	57	57	53	52	53	53	55	105	105	131	134	134	157	157	193	193	180	180	238	238	285	285	345	345	323	323				
#7	113	113	113	110	107	108	108	104	103	103	127	129	129	148	148	178	178	185	185	206	206	261	261	309	309	311	311				
#6	127	127	127	130	129	133	133	138	133	133	127	127	127	150	150	181	181	178	178	205	205	195	195	66	66	59	59				
#5	124	124	124	125	129	133	133	134	139	139	150	150	150	157	157	154	154	152	152	109	109	15	15								
#4	105	105	105	107	105	100	100	98	117	117	116	112	112	106	106	60	60	56	56	6	6										
#3	97	97	97	99	102	102	102	100	86	86	61	60	60	29	29	2	2														
#2	0	0	0	0	0	0	0	25	26	26	20	19	19																		
#1	24	24	24	24	25	25	25																								

Fig. 11.28 The spatial distribution of vehicle counts in the toll plaza.

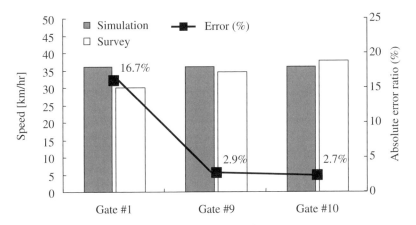

Fig. 11.29 A comparison of average vehicle speeds at the ETC toll gate.

Gate	0m	10m	20m	30m	40m	50m	60m	70m	80m	90m	100m	110m	120m	130m	140m	150m	160m	170m	180m	190m	200m	210m	220m	230m	240m	250m	260m	270m	280m	290m	300m
#11	10	20	31	36	41	46	46	51	57	57	62	67	67	72	72																
#10	36	36	36	44	45	52	52	51	57	57	64	64	68	72	72	72	72	72	72	72	72	72	72	72	72	72					
#9	36	37	37	43	45	52	52	54	57	57	60	60	68	71	71	71	71	71	71	71	70	70	64	64	71	71	71	71	72	72	
#8	10	20	30	36	41	46	46	50	54	54	59	65	65	72	72	71	71	72	72	70	70	69	69	67	67	70	70	69	69	71	71
#7	6	18	28	33	40	45	45	51	54	54	59	63	63	68	68	69	69	72	72	68	68	67	67	70	70	67	67	69	69	71	71
#6	6	19	29	36	40	45	45	50	52	52	57	66	66	72	72	70	70	70	70	70	70	70	72	72	72	72	72				
#5	8	19	28	35	41	45	45	51	56	56	58	64	64	72	72	72	72	72	72	72	72	72									
#4	6	19	28	33	36	44	44	51	56	56	58	67	67	72	72	72	72	72	72	72											
#3	8	18	28	36	40	43	43	51	50	50	58	67	67	72	72	72	72														
#2	10	20	31	36	41	46	46	49	57	57	62	67	67																		
#1	36	40	40	42	48	52	52																								

Fig. 11.30 The spatial distribution of the average running speed in the toll plaza.

11.4.2   Validation of speed

Figure 11.29 compares the average vehicle speeds at the ETC toll gates. Since the free flow speed at the toll gate position was given as a parameter, it is reasonable for the simulation to reproduce vehicle speeds at free flow conditions. Figure 11.30 shows the spatial distribution of the average speed every 10 meters on each lane. The unit of the numbers in the gray scale cells is [km/hr], and a darker color refers to slower speed sections. In this case of light traffic, no slow-down sections were found except for near the non-ETC gates.

11.4.3   Count of 'near-miss' opportunities

One of the major purposes of developing this simulation model was to assess the safety of the toll plaza design. It was, however, quite difficult to evaluate the safety issues with traffic simulation by directly reproducing 'dangerous' situations, since such extreme and rare phenomena are rarely observed and hard to model. The Monte Carlo Method, like this traffic simulation, is not capable of dealing with such rare phenomena.

For the purpose of evaluating the traffic safety, we took into account Heinrich's law, which states that a serious accident or event would be preceded by a number of

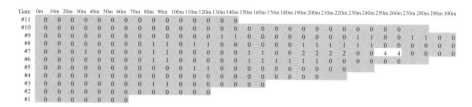

Fig. 11.31 The spatial distribution of 'near-miss' counts in the toll plaza.

similar 'near-miss' events. Here, we thus invented 'near-miss' opportunities according to the following conditions.

(a) Two vehicles, both at free flow speed, were driving on two adjacent lanes;

(b) One of them started a lane change and squeezed himself in front of the other with the headway of less than 1.0 second.

Surely this situation cannot be regarded as immediately dangerous, but it may represent some extent of the pressure that a driver may feel. As for the threshold headway of 1.0 second, it tentatively produces a much higher flow rate than the road capacity, forcing the following driver to decelerate against his/her will. This can be regarded as an 'unpleasant' or 'near-miss' situation. Let us consider the number of such 'near-miss' counts at a certain position to obtain a degree of the safety in that place. Note that the 'near-miss' events will not be counted when the vehicles are in a congested section, according to condition a) above.

Figure 11.31 illustrates the spatial distribution of 'near-miss' counts. Although these numbers are small because of the light traffic condition in the validation case, one can observe a small tendency of 'near-miss' events taking place to a larger extent at upstream sections of the toll plaza.

11.5 CASE STUDIES

For the purpose of planning an optimal design of a toll plaza with a high flow of ETC vehicles, a number of case studies were executed. This chapter presents the results of three case studies listed in Table 11.2.

The traffic demand was set to the level of the 30th hourly traffic volume from annual statistics at the Narashino Toll Plaza. The portion of ETC vehicles was set to 80%, i.e., about 2400 ETC vehicles arriving per hour.

Since the capacity of the ETC toll gate was estimated at 800 [veh/hr] in the video survey, three ETC exclusive lanes, as in the current configuration, would be marginal for the demand. Under heavy traffic conditions, many lane changes may be carried out, thus affecting the capacity of the toll plaza. Therefore, in Case-80-2 and Case-80-3, one more ETC exclusive lane was added to the current lane configuration.

Table 11.2 The settings for the case studies.

Case	Traffic demand	ETC vehicles	Exclusive ETC gates
Case-80-1	3100 [veh/hr]	80%	Gate #1, #9, #10
Case-80-2	3100 [veh/hr]	80%	Gate #1, #2, #9, #10
Case-80-3	3100 [veh/hr]	80%	Gate #1, #9, #10, #11

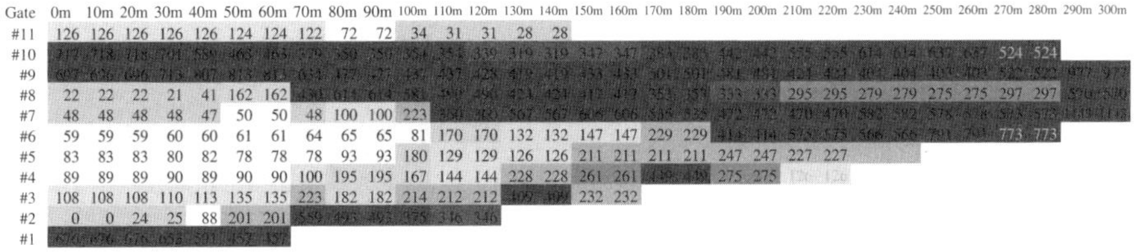

Gate	0m	10m	20m	30m	40m	50m	60m	70m	80m	90m	100m	110m	120m	130m	140m	150m	160m	170m	180m	190m	200m	210m	220m	230m	240m	250m	260m	270m	280m	290m	300m
#11	126	126	126	126	126	124	124	122	72	72	34	31	31	28	28																
#10																												524	524		
#9																															
#8	22	22	22	21	41	162	162															295	295	279	279	275	275	297	297		
#7	48	48	48	48	47	50	50	48	100	100	223																				
#6	59	59	59	60	60	61	61	64	65	65	81	170	170	132	132	147	147	229	229									773	773		
#5	83	83	83	80	82	78	78	78	93	93	180	129	129	126	126	211	211	211	211	247	247	227	227								
#4	89	89	89	90	89	90	90	100	195	195	167	144	144	228	228	261	261			275	275										
#3	108	108	108	110	113	135	135	223	182	182	214	212	212			232	232														
#2	0	0	24	25	88	201	201																								
#1																															

Fig. 11.32 The spatial distribution of vehicle counts in the toll plaza (Case-80-1).

Gate	0m	10m	20m	30m	40m	50m	60m	70m	80m	90m	100m	110m	120m	130m	140m	150m	160m	170m	180m	190m	200m	210m	220m	230m	240m	250m	260m	270m	280m	290m	300m
#11	6	18	29	35	40	45	45	50	51	51	62	67	67	70	70																
#10	33	25	25	20	15	16	16	24	32	32	40	40	51	48	48	50	50		47	24	24	14	14	13	13	13	13				
#9	35	28	28	27	21	13	13	7	5	6	6	6	8	9	9		7	7	6	5	5	6	8					26	26		
#8	10	20	31	36	23	16	16	12	10	10	8	6	6	5	5	5	4	4	4	4	4	5				11	11			14	14
#7	10	20	30	34	41	45	45	47	21	21	15	16	16	15	15	12	12		8	8		11	11	10	10	13	13	38	38		
#6	9	19	29	35	40	45	45	51	53	53	62	63	63	67	67	37	37	25	25	22	22	25	25	29	29	44	44				
#5	9	19	30	35	41	45	45	51	57	58	57	57	63	63	64	64	66	66	58	58	71	71									
#4		18	28	35	40	45	45	48	45	45	44	45	45	61	61	57	57	63	63	64	64										
#3	6	19	29	35	40	36	36	35	27	27	25	34	34	53	53	59	59														
#2	10	20	7	32	11	12	12	23	19	19	20	31	31																		
#1	36	22	22	20	14	15	15																								

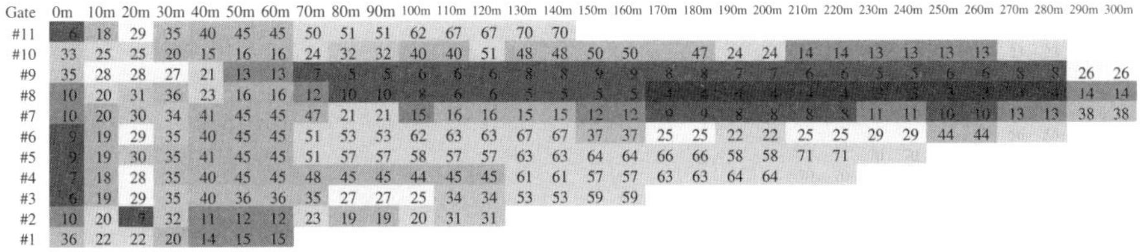

Fig. 11.33 The spatial distribution of vehicle speeds in the toll plaza (Case-80-1).

11.5.1 Case-80-1

In this case, the traffic at the toll plaza is jammed, despite that the ETC toll gates (#8, #0 and #1) are not operating at full capacity, as shown in Figure 11.32. From the speed distribution in Figure 11.33, it was found that the section from 160 m to 220 m is the one that is slowed down the most. This implies that the interference between vehicles changing lanes declines the capacity of this section less than that of the ETC gate. As a result, it can be concluded that three ETC lanes are not enough for a case with 80% ETC vehicles.

Figure 11.34 shows the number of 'near-miss' events. Since a 'near-miss' was defined to count only in free flow condition and the present case referred to a traffic jam, fewer 'near-miss' opportunities would be recorded.

11.5.2 Case-80-2

Figures 11.35 and 11.36 show the vehicle counts as well as the speeds on each lane. In this case, the traffic jam in the toll plaza is less severe than for Case-80-1 but the speed of the vehicles has been slightly reduced, in spite of the ETC gates not being utilized to their full capacity. In particular, the speed of the section 60 m to 100 m was slower than the speed at the gate, which implies that the lane changes in this section affected the flow of traffic. The 'near-miss' counts in Figure 11.37 support the idea that there

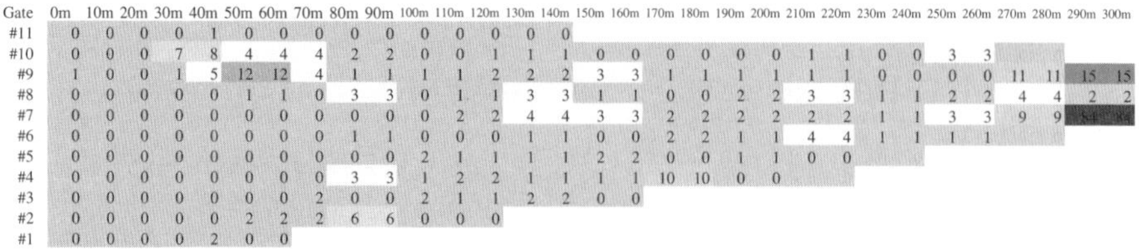

Gate	0m	10m	20m	30m	40m	50m	60m	70m	80m	90m	100m	110m	120m	130m	140m	150m	160m	170m	180m	190m	200m	210m	220m	230m	240m	250m	260m	270m	280m	290m	300m
#11	0	0	0	0	1	0	0	0	0	0	0	0	0	0	0																
#10	0	0	0	7	8	4	4	4	2	2	0	0	1	1	1	0	0	0	0	0	1	1	0	0	3	3					
#9	1	0	0	1	5	12	12	4	1	1	1	1	2	2	2	3	3	1	1	1	1	1	1	0	0	0	0	11	11	15	15
#8	0	0	0	0	0	1	1	0	3	3	0	1	1	3	3	1	1	0	0	2	2	3	3	1	1	2	2	4	4	2	2
#7	0	0	0	0	0	0	0	0	0	0	0	2	2	4	4	3	3	2	2	2	2	2	2	1	1	3	3	9	9		
#6	0	0	0	0	0	0	0	0	1	1	0	0	0	1	1	0	0	2	2	1	1	4	4	1	1	1	1				
#5	0	0	0	0	0	0	0	2	1	1	1	1	2	2	0	0	1	1	0	0											
#4	0	0	0	0	0	0	0	3	3	1	2	2	1	1	1	1	10	10	0	0											
#3	0	0	0	0	0	0	0	2	0	0	2	1	1	2	2	0	0														
#2	0	0	0	0	0	2	2	2	6	6	0	0	0																		
#1	0	0	0	0	2	0	0																								

Fig. 11.34 The spatial distribution of 'near-miss' counts in the toll plaza (Case-80-1).

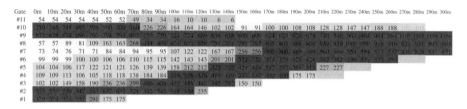

Fig. 11.35 The spatial distribution of vehicle counts in the toll plaza (Case-80-2).

Gate	0m	10m	20m	30m	40m	50m	60m	70m	80m	90m	100m	110m	120m	130m	140m	150m	160m	170m	180m	190m	200m	210m	220m	230m	240m	250m	260m	270m	280m	290m	300m
#11		19	30	36	40	45	45	50	55	55	62	67	67	72	72																
#10	35	28	28	21	20	23	23	29	39	39	47	47	61	62	68	68	69	69	71	71	68	68	62	62	60	60					
#9	36	27	27	26	24	22	22	19	19	19	20	20	23	31	31	43	43	53	53	55	55	55	55	52	52	58	58	63	63	60	60
#8		19	19	34	23	24	24	24	27	27	35	41	41	43	43	45	45	47	47	43	43	40	40	38	38	37	37	49	49	60	60
#7		19	30	34	38	40	40	43	47	47	53	58	58	58	58	60	60	59	59	57	57	54	54	57	57	59	59	65	65	67	67
#6		19	28	34	37	43	43	51	50	50	60	61	61	61	61	63	63	60	60	59	59	61	61	67	67	71	71				
#5		19	29	32	39	44	44	50	50	50	52	53	53	57	57	57	57	60	60	64	64	70	70								
#4		19	30	33	37	40	40	39	35	35	39	45	45	56	56	62	62	65	65	68	68										
#3		19	11	23	23	25	25	25	23	23	32	45	45	62	62	67	67														
#2	12	13	13	14	20	24	24	23	19	19	23	23	47																		
#1	35	33	33	31	33	39	39																								

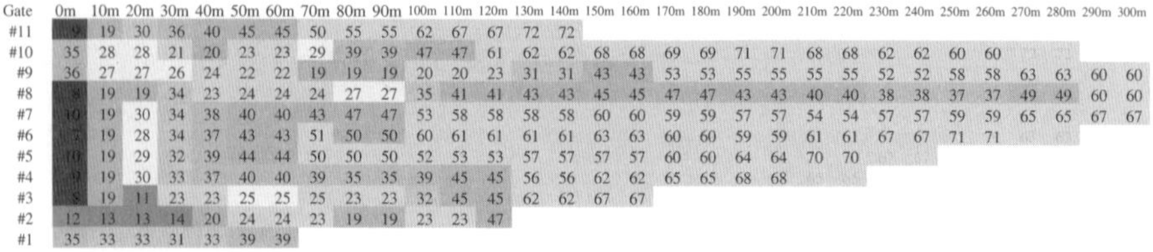

Fig. 11.36 The spatial distribution of vehicle speeds in the toll plaza (Case-80-2).

Fig. 11.37 The spatial distribution of 'near-miss' counts in the toll plaza (Case-80-2).

are many 'near-miss' events along this section, as well as in the upstream section around the 200-m position. A similar situation with regard to 'near-miss' counts is found on the left side of the toll plaza.

11.5.3   Case-80-3

Figures 11.38 and 11.39 show the vehicle counts and speeds on each lane, and as for the other two cases, the ETC gates did not operate at their full capacities. The slow section on the ETC lane was longer in this case than in Case-80-2, but never extended outside the toll plaza section. As shown in Figure 11.40, the 'near-miss' counts were fewer than in Case-80-2, as less counts were recorded in the left side of the toll plaza.

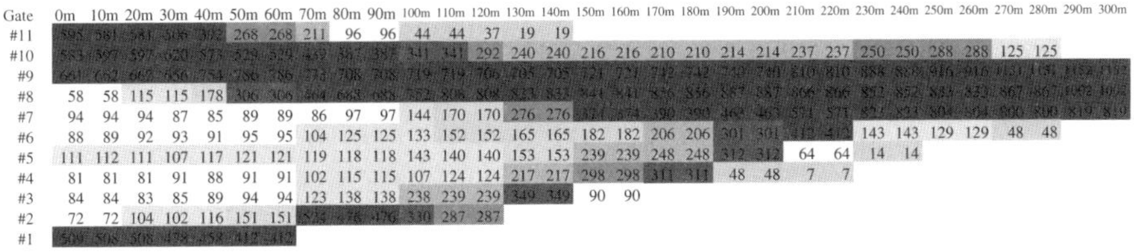

Gate	0m	10m	20m	30m	40m	50m	60m	70m	80m	90m	100m	110m	120m	130m	140m	150m	160m	170m	180m	190m	200m	210m	220m	230m	240m	250m	260m	270m	280m	290m	300m
#11	595	581	541	508	307	268	268	211	96	96	44	44	37	19	19																
#10	583	567	597	620	573	529	520	451	387	341	292	240	240	216	216	210	210	214	214	237	237	250	250	288	288	125	125				
#9	660	652	665	656	753	786	786	771	708	708	719	714	706	705	721	721	732	742	740	748	810	810	888	870	916	916	1031	1036	1059		
#8	58	58	115	115	178	306	306	364	663	683	762	808	823	833	841	841	856	856	857	857	866	866	857	852	855	833	867	867	1052	1054	
#7	94	94	94	87	85	89	89	86	97	97	144	170	170	276	276		90	393	465	465	571	571	82	833	809	809	819	819			
#6	88	89	92	93	91	95	95	104	125	125	133	152	152	165	165	182	182	206	206	301	301			143	143	129	129	48	48		
#5	111	112	111	107	117	121	121	119	118	118	143	140	140	153	153	239	239	248	248	312	311	64	64	14	14						
#4	81	81	81	91	88	91	91	102	115	115	107	124	124	217	217	298	298	311	311	48	48	7	7								
#3	84	84	83	85	89	94	94	123	138	138	238	239	239	349	349	90	90														
#2	72	72	104	102	116	151	151	818	678	476	330	287	287																		
#1	509	508	508	478	358	317	310																								

Fig. 11.38 The spatial distribution of vehicle counts in the toll plaza (Case-80-3).

Gate	0m	10m	20m	30m	40m	50m	60m	70m	80m	90m	100m	110m	120m	130m	140m	150m	160m	170m	180m	190m	200m	210m	220m	230m	240m	250m	260m	270m	280m	290m	300m
#11	15	16	16	17	22	29	29	43	51	51	63	63	66	72	72																
#10	14	17	17	18	19	23	23	29	43	43	45	45	56	57	57	61	61	63	63	63	63	55	55	57	57	55	55	69	69		
#9	35	25	25	22	19	15	15	12	11	11	12	12	13	15	15	17	17	21	21	23	23	27	27	32	32	52	52	62	62	56	56
#8	10	20	19	35	22	22	22	22	21	21	22	22	21	21	22	23	23	24	24	23	23	24	24	29	29	45	45	60	60		
#7	10	20	30	34	41	45	45	44	46	46	43	43	43	46	46	48	48	59	59	53	53	52	52	56	56	54	54	58	58	64	64
#6	78	19	29	34	40	44	44	50	51	51	59	63	63	71	71	69	69	59	59	61	61	66	66	61	61	72	72	72	72		
#5	9	20	28	32	40	45	45	45	53	53	58	64	64	68	68	65	65	64	64	71	71	72	72	72	72						
#4	10	19	29	34	38	43	43	48	52	52	57	55	55	67	67	68	68	69	69	65	65	72	72								
#3	10	19	29	33	38	40	40	43	35	35	39	56	56	67	67	70	70														
#2	9	19	18	31	23	28	28	39	34	34	35	55	55																		
#1	36	26	26	25	26	35	35																								

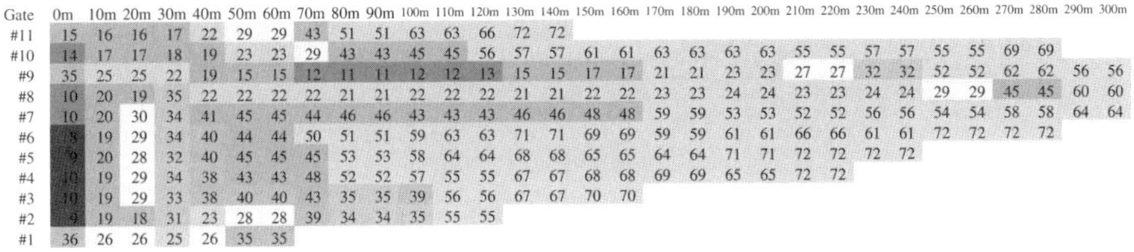

Fig. 11.39 The spatial distribution of vehicle speeds in the toll plaza (Case-80-3).

Gate	0m	10m	20m	30m	40m	50m	60m	70m	80m	90m	100m	110m	120m	130m	140m	150m	160m	170m	180m	190m	200m	210m	220m	230m	240m	250m	260m	270m	280m	290m	300m
#11	1	6	6	10	5	1	1	3	0	0	0	0	0	0	0																
#10	1	6	6	16	9	6	6	5	4	4	3	3	2	1	1	1	1	0	0	1	1	1	1	0	0	6	6	0	0		
#9	3	5	5	8	12	16	16	14	7	7	5	5	5	4	4	4	6	6	3	3	6	6	3	3	2	2				3	3
#8	0	0	0	0	0	0	0	1	3	3	1	5	5	6	6	3	10	10	11	11	16	16	5	5	11	11	16	16	10	10	
#7	0	0	0	1	0	1	1	1	0	0	0	0	0	1	3	3	3	3	5	5	8	8	8	8	11	11	8	8	8	8	
#6	0	0	0	1	0	0	0	0	1	1	0	1	0	1	1	3	3	4	4	4	2	2	0	0	0	0					
#5	0	0	1	0	0	0	0	1	0	0	2	1	1	0	0	3	3	1	1	0	0	0	0								
#4	0	0	0	1	0	0	0	0	0	0	0	1	1	0	0	0	1	1	0	0	0	0									
#3	0	0	0	1	0	0	1	0	0	2	2	1	1	0	0																
#2	0	0	0	0	0	1	1	4	3	3	3	2	0	0																	
#1	1	0	0	0	2	0	0																								

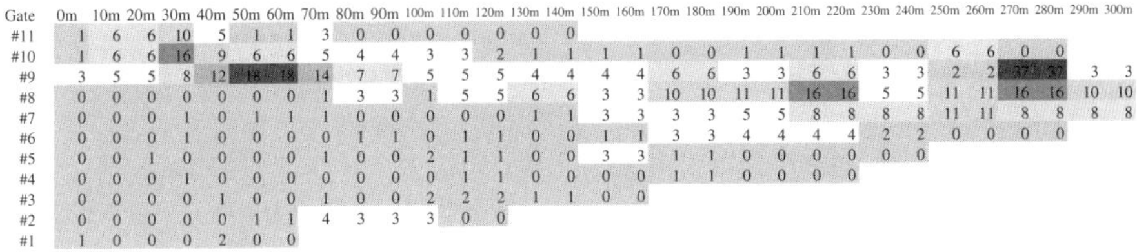

Fig. 11.40 The spatial distribution of 'near-miss' counts in the toll plaza (Case-80-3).

Table 11.3 The summary of the case studies.

	Case-80-1	Case-80-2	Case-80-3
Total travel time [veh.hr]	72.48	30.19	36.72
Average TT for all vehicles [s]	74.7	42.4	43.7
for ETC vehicles [s]	77.8	36.7	38.8
for non-ETC vehicles [s]	73.3	44.8	45.8
Counts of 'near-miss'	585	926	786

11.5.4 Summary of the case studies

The results of the three case studies are summarized in Table 11.3. Regarding the efficiency, Case-80-2 showed the best result. However, the results from Case-80-3 comprised fewer 'near-miss' events than Case-80-2, suggesting that there might be less pressure on the drivers. Since the average travel times in Case-80-3 were almost the same as in Case-80-2, this case might be considered further for the future plan.

The case studies provided several useful implications for the improvement of the toll plaza design and operation, e.g.:

(1) Bottlenecks should not appear at the ETC gate but rather at a position 50 to 100 meters upstream from the ETC gate due to the merging of ETC vehicles.

(2) Near-miss event increased not only at the bottleneck position near the toll gate but also farther from the gate where the congested section reached.

(3) If another ETC gate were to be added to improve the capacity of the toll plaza, it would be better to replace one of the normal gates near the center rather than one on the left side, as this should give rise to fewer 'near-miss' counts.

11.6   CONCLUSIONS

The present chapter describes the development of a simulation model for the evaluation of the design of an ETC toll plaza. Prior to the modeling stage, a precise video imaging survey was conducted at the Narashino Toll Plaza on the Higashi-Kanto Expressway. The simulation model was designed based on the findings from this survey, by utilizing data from normal traffic, such as speeds, headways, lane choice, etc. The flow model and the lane choice model proposed herein were calibrated and validated by applying them to the survey results. Case studies for situations with high flows of ETC vehicles were evaluated, and important implications were obtained through the simulation application.

The present chapter only explains case studies for toll barriers, and thus the model parameters identified here may not be universal. Further studies on different types of toll plazas are envisioned as future work.

11.7   REFERENCES

Akahane, H., and Hatakenaka, S. (2004) "Successive observations of trajectories of vehicles with plural video cameras", *International Journal of ITS Research*, 2(1).
Horiguchi, R., Kuwahara, M., Katakura, M., Akahane, H., and Ozaki, H. (1996) "A network simulation model for impact studies of traffic management 'AVENUE Ver. 2'", *Proceedings of 3rd World Congress on Intelligent Transport Systems*, Orlando (CD-ROM).
Yoshii, T., and Kuwahara, M. (1995) "A traffic simulation model for oversaturated traffic flow on urban expressways", *Preprint at 7th World Conference on Transportation Research*, Sydney.

CHAPTER 12

TIME-DEPENDENT ORIGIN-DESTINATION ESTIMATION WITHOUT ASSIGNMENT MATRICES

Ramachandran Balakrishna, Moshe Ben-Akiva, Haris N. Koutsopoulos

Time-dependent origin-destination (OD) flows are crucial inputs to dynamic traffic assignment (DTA) models. However, they are often unobserved, and must be estimated from indirect traffic measurements collected from the study network. Approaches to estimate OD flows from link counts traditionally rely on assignment matrices that map the OD flow variables onto the counts. However, this method (a) approximates the complex relationship between OD flows and counts with a linear function, (b) is restricted to the use of only counts, and cannot exploit richer data such as speeds, densities or travel times, and (c) cannot estimate route choice and supply parameters that critically impact the OD estimates. This chapter presents a dynamic OD estimation method that is accurate and flexible in the use of general traffic data. Moreover, it simultaneously estimates all parameters with an impact on OD estimation, and can be applied to any traffic assignment model.

12.1 INTRODUCTION

Dynamic Traffic Assignment (DTA) models are used to estimate and predict traffic conditions in urban networks through complex demand and supply model components that interact systematically to simulate the performance of the network. Such models are designed for a variety of transportation planning, operations and management applications. The current state-of-the-art DTA models are simulation-based, such as DynaMIT (Ben-Akiva et al., 2001; DYNASMART Mahmassani, 2002). Such models estimate and predict traffic conditions, including flows, densities and travel times, using detailed models of travel demand and network supply, as well as sensor data.

The structure of a generic DTA model is outlined in Figure 12.1. Demand model components estimate and predict time-varying (dynamic) OD trip patterns, and simulate the behavior of individual motorists (including pre-trip departure time,

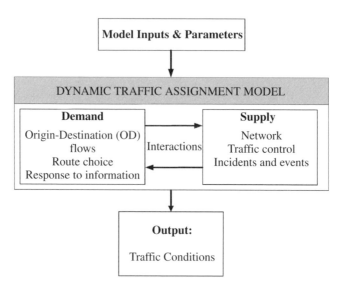

Fig. 12.1 The structure of a generic DTA.

mode and route choices, and response to information). Supply model components capture traffic phenomena through detailed representations of the capacities of network elements, the traffic dynamics resulting from speed/acceleration, lane-changing and merging/weaving behavior, as well as the impact of incidents. Various algorithms address the interactions between demand and supply to assign the dynamic demand to the network and determine the temporal propagation of flows. The resulting traffic conditions (including speeds, densities, travel times and delays) may be used for a variety of planning and real-time management applications.

A key input to any DTA model is the dynamic OD profile, generally represented by a series of OD matrices of trip rates between origin and destination points. Each matrix specifies the trips that depart from their origins during a specific time interval. However, such detailed demand data are typically unobserved. Existing demand matrices are often static (and measure average flows across extended time periods). Moreover, the difficulties associated with conducting OD surveys render many such matrices outdated.

Initial attempts at OD estimation using vehicle count data have focused on identifying a matrix of *static* OD flows (see for example Cascetta, 1984; Cascetta and Nguyen, 1988; Hazelton, 2000, 2001; Yang et al., 1992). These approaches estimate a matrix of average OD flows (assumed to be constant over a significant time period such as the entire morning peak), based on counts collected for the same period on a subset of network links. While static OD flows may suffice for several long-term planning applications, a finer resolution is desired for within-day and short-term operations and management problems. The recent deployment of Intelligent Transportation Systems has resulted in large databases containing rich, time-varying traffic data such as counts, speeds and densities. Such data are collected and archived automatically, so that data on the most recent traffic conditions can be collected regularly and at a reasonable cost. The availability of time-varying data provides the opportunity to estimate OD flow matrices by time of day, suitable for DTA applications.

Several approaches to dynamic OD estimation have been developed, which divide the study period into multiple time intervals, $h = 1,2,...,H$. A matrix x_h of trips departing during interval h is estimated based on counts measured during interval y_h and beyond. Cascetta et al. (1993) proposed a statistical updating framework in which target OD flows were adjusted to better replicate the observed count data. An optimization step was performed, subject to non-negativity constraints on the OD flows:

$$\hat{x} = \arg\min[z_1(x,x^a) + z_2(f(x),y)] \tag{12.1}$$

where

x : $\begin{bmatrix} x_1 \cdots x_h \cdots x_H \end{bmatrix}$ is the matrix of OD vectors to be estimated;

$\hat{x}$: the required optimal estimates;

x^a : the target OD flows;

y : $\begin{bmatrix} y_1 \cdots y_h \cdots y_H \end{bmatrix}$ are traffic counts;

z_1 and z_2 : goodness-of-fit measures (e.g., least squares).

The function $f(\ )$ provides the mapping between the unknown OD flows and the counts, and is a complex function of the OD demand, driver route choice decisions and the resultant network travel times. It is typically approximated with a linear assignment matrix mapping with the following structure through an autoregressive process of degree q':

$$y_h = \sum_{p=h-p'}^{h} a_h^p x_p + v_h \tag{12.2}$$

where

y_h : vector of link sensor counts for interval h;

x_p : vector of OD flows departing from their origins during time interval p;

a_h^p : assignment matrix;

v_h : error vector.

p' : number of intervals spanning the longest trip.

Ashok and Ben-Akiva (2000) developed a state-space model for the real-time estimation of OD flows. This approach employs a measurement equation similar to Eq. (12.2) together with a transition equation that captures the evolution of system *state*:

$$x_h - x_h^H = \sum_{p=h-q'}^{h} f_h^p \left(x_p - x_p^H \right) + w_h \tag{12.3}$$

The term w_h is an error vector. A key innovation in the above equation is the definition of state in terms of *deviations* of OD flows x_h from their historical values x_h^H. The deviations aid in efficiently including all information in prior intervals while estimating flows for the current interval h. Further, deviations may be modeled by symmetrical distributions such as the normal. The model is sequentially solved using a Kalman Filter, with historical flows initialized through a least squares procedure.

Long, multi-interval trips may be captured in the above method through state augmentation, so that a moving set of flows in past intervals is re-estimated at each step. Ashok (1996) presented an off-line extension to his real-time estimator, in which the sequential OD estimates in a forward pass of the Kalman Filter were "smoothed" during a backtracking step. The updated OD flows for interval h thus reflected count information from future intervals.

Both types of approaches outlined above require knowledge of the assignment matrices a_h^p. Assignment fractions can be calculated easily under free-flow conditions, or when travel times on all links are known (see van der Zijpp, 1996). However, if the network is congested, the assignment fractions become functions of prevailing travel times, which, in turn, depend on the OD flows that are yet to be determined. A fixed-point problem is thus obtained.

Cascetta and Postorino (2001) presented various algorithmic approaches to solve for a fixed point in the static context. They searched for consistent OD flows and assignment fractions by iterating between the OD estimation step and a network loading method. The iterative nature of the solution, while still a heuristic, also possessed high computational overhead.

Assignment matrix-based dynamic OD estimation formulations are also characterized by computational limitations, as discussed below:

- The assignment matrix is a linear approximation of the complex relationships between OD flows and sensor counts, which might only be valid close to the optimal OD flows. In general, though, an iterative fixed-point problem must be solved. Existing solution approaches are generally based on heuristics.

- Since the computational effort of simultaneously estimating OD flows for multiple time intervals involves the calculation, storage and inversion of a large, augmented assignment matrix, it has been found to be prohibitive even on medium-sized networks, (see Bierlaire and Crittin, 2004; Cascetta and Russo, 1997; Toledo et al., 2003). Sequential OD estimation (see for example Cascetta et al., 1993) has therefore been used as an approximation that fixes the OD flows departing in all prior intervals. However, counts from future intervals are not used to refine these past estimates, which can affect the accuracy of OD estimates on large, congested networks with trips spanning many intervals.

- Relationships between general data (such as speeds and densities) and OD flows are expected to be non-linear, and approximations similar to the assignment matrix cannot be justified. The assignment matrix formulation thus restricts OD estimation to the use of count data alone, which can potentially over-fit to counts at the expense of traffic dynamics.

- The assignment matrix excludes the estimation of other related parameters such as route choice or supply models, together with OD flows. These parameters can significantly impact the calculation of the assignment matrix (see Park et al., 2005).

As a result, the popular assignment matrix formulation possesses several drawbacks. This chapter presents an estimation method addressing the above limitations in existing approaches. The remainder of the chapter is organized as follows: The proposed OD estimation methodology is discussed, including a general mathematical formulation and efficient solution algorithms. Two case studies are presented to illustrate accuracy and scalability. The chapter concludes with a summary of the proposed methodology.

12.2 METHODOLOGY

A systematic OD estimation approach that does not rely on the calculation of assignment matrices is developed. Rather, the complex relationships between the OD flows and traffic measurements are captured directly, by treating the assignment model as a black box. The problem may be formulated mathematically in the following optimization framework:

$$\underset{x_1,\ldots,x_H}{\text{Minimize}} \sum_{h=1}^{H} \left[z_1\left(M_h, \mathcal{M}_h\right) + z_2\left(x_h, x_h^a\right) \right] \qquad (12.4)$$

subject to the following constraints:

$$\mathcal{M}_h = f\left(x_1,\ldots,x_h; \beta; G_1,\ldots,G_h\right) \ \ \forall \ \ h \in \{1,2,\ldots,H\} \qquad (12.5)$$

where

M_h and $\mathcal{M}_h$	: observed and fitted sensor measurements for time interval h;
x_h	: vector of OD flows departing during interval h;
x_h^a	: corresponding *a priori* values;
z_1 and z_2	: goodness-of-fit functions;
$f(\bullet)$	: DTA model output.

$\mathcal{M}_h$ represents the DTA model, and includes both route choice and network loading components. The model output (e.g., counts) for interval h is a function of the evolution of traffic up to interval h (in other words, demand entering the network in future intervals does not influence the counts in the current or prior intervals).

Besides the OD flows, the fitted measurements may also be functions of other model parameters β such as the coefficients in a route choice model, segment capacities and link performance functions. Moreover, the DTA output depends on the network G_h (for example, reversible high occupancy vehicle (rHOV) links may only be available at specific times of day). The assignment step involves the calculation of route choice probabilities, the conversion of OD flows to path flows, and the loading of path flows into the network, and is influenced by all these inputs and parameters.

The direct use of the assignment model's output provides several benefits since it:

- Accurately captures the complex nonlinear dependencies between the OD variables x_h and the data, without the use of linear approximations;

- Allows the search space to be augmented with variables other than OD flows. Supply and route choice parameters (denoted by the vector β) critically impact the calculation of the assignment matrix, yet their relationships to the OD flows are often complex and non-linear. They have therefore been treated by traditional OD estimation as exogenous inputs, which can result in biased flows. The simultaneous determination of all such parameters (together with OD flows) is therefore desirable. The proposed approach allows for x_h and β to be consistently and simultaneously estimated, using a common set of traffic measurements, through a simple modification to Eq. (12.4):

$$\text{Minimize}_{x_1,\dots,x_H,\beta} \sum_{h=1}^{H} \left[z_1\left(M_h, \mathcal{M}_h\right) + z_2\left(x_h, x_h^a\right) \right] + z_3\left(\beta, \beta^a\right) \tag{12.6}$$

where β^a is a vector of *a priori* model parameter values, and z_3 is a goodness-of-fit function. The constraints in Eq. (12.5) remain unaltered. The new formulation

- Introduces the flexibility to easily include any general traffic measurements, beyond standard link counts. It is believed that OD estimation based only on counts may lead to the identification of just one of several possible solutions. Information such as speeds or densities, also recorded by many loop detectors, may thus be vital in selecting a solution that also matches traffic dynamics. In addition, more advanced indirect measurements are becoming available, including point-to-point counts and travel times from automatic vehicle identification (AVI) systems and probe vehicle travel times. However, a linear mapping for these data types is harder to compute. Such data, typically excluded from the OD estimation process so far, may now be incorporated efficiently.

  The relationships between OD matrices and general data (such as speeds, densities, etc.) are very complex and non-linear. In our methodology, we obtain the output of these relationships directly from the simulator, through the function $\mathcal{M}_h$. Instead of assuming a specific analytic relationship between OD flows and for example, speeds, the simulation is used to obtain this relation. Thus, the simulation acts as a black box function. For instance, the speeds at sensors are outputs of running the simulator with a given set of OD inputs, and can be compared against real-world speed measurements. Therefore, the proposed approach does not require the analytical form of these relationships to be known, and this is a significant strength of the methodology.

- Becomes applicable to any traffic assignment, including analytical models and microscopic, mesoscopic and macroscopic simulation tools.

Together, Eq. (12.4) and (12.5) represent a complex, non-linear, non-analytical optimization problem, due to the use of a sophisticated simulator to obtain the fitted measurements. When the simulator is stochastic (as is often the case with realistic models), the optimization problem must also account for the inherent noise in model outputs. Further, the problem is large in scale, with the number of OD pairs and time intervals increasing rapidly with the size of the network and the desired temporal modeling resolution. Consequently, appropriate solution algorithms must be able to:

- move through possible local optima to locate global solutions;

- work without analytical derivatives, which are generally unavailable;

- have reasonable computational performance and be applicable to large networks.

Population-based simulation optimization methods such as Box-Complex (see Box, 1965) and Stable Noisy Optimization by Branch and Fit (SNOBFIT) (see Huyer and Neumaier, 2004) are ideally suited for such applications. While their capabilities have been demonstrated in other areas, their application to complex and large-scale transportation problems is limited. The present study proposes a solution approach that combines these two non-linear simulation optimization algorithms. The Box-Complex algorithm is first used to randomly span the feasible space and identify a set of points around potential local minima. SNOBFIT is subsequently employed to perform local optimization around each point, through quadratic approximation. The two-stage process ensures that the detailed local searches take place around points with a high probability of optimality. The SPSA algorithm (see Spall, 1992) has also performed well in recent experiments, both in terms of solution quality and computational efficiency (see Balakrishna, 2006). SPSA requires just two objective function evaluations per iteration, irrespective of the size p of the calibration problem (i.e., the number of unknown OD flows). In contrast, traditional gradient-based approaches must evaluate $2p$ objective functions per iteration. Since each objective function evaluation requires a costly simulation run, it is easy to perceive the computational efficiency of SPSA for large-scale problems.

Here, the feasibility and advantages of the proposed methodology are demonstrated through two practical case studies using a state-of-the-art traffic simulation model.

12.3   CASE STUDY I: SYNTHETIC NETWORK

The objectives of this case study are (a) to demonstrate the feasibility and accuracy of the OD estimation methodology developed in the previous section, and (b) to illustrate the ability of the proposed approach to replicate known OD patterns underlying the observed traffic data.

The test network (Fig. 12.2) was a directed graph consisting of 8 nodes and 8 links. Each link was divided into 3 segments in order to better represent the spatial and temporal evolution of congestion. Demand was assumed to flow between all three feasible OD pairs (connecting the three origin nodes O1, O2 and O3 to the destination node D). Travelers making trips between O1 and D could choose from two alternative paths, while the remaining two OD pairs were captive to a single path each. Link geometries reflect a varying numbers of lanes, as indicated in the figure.

The period of interest spanned 50 minutes, designated arbitrarily as 6:50 AM–7:40 AM. This time period was further divided into uniform intervals, each with a duration of 5 minutes (so that the number of intervals $H = 10$). An incident was assumed to impact the available capacity on the network for a period of 10 minutes, beginning at 7:05 AM, and the location of this incident is indicated in Figure 12.2. One out of two available lanes at the incident location was blocked while the incident was in effect. Drivers from origin O1 who chose the path affected by the incident were thus likely to incur further delays.

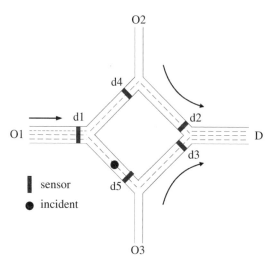

Fig. 12.2 The synthetic test network.

MITSIMLab (Yang and Koutsopoulos, 1996; Yang et al., 2000), a detailed microscopic traffic simulation laboratory, was used to represent the real world and the corresponding "true" state of the network. MITSIMLab simulates the movement of individual vehicles between various OD pairs on the "real" traffic network through detailed models of driving behavior and route choice decision-making. The system also represents a variety of traffic management strategies, and captures the operations of traffic control systems and the impact of incidents on network supply.

MITSIMLab possesses several unique features (Balakrishna et al., 2005) motivating its use for generating synthetic sensor data for the test network. The system can accommodate a wide range of assumptions related to the generation of network demand, travel and driving behavior (and resultant traffic dynamics) and incidents. In addition, MITSIMLab replicates the operation of the surveillance (sensor) system, providing a flexible setting that can mimic traffic data under diverse demand and traffic situations. It should be pointed out that the modeling of traffic dynamics in MITSIMLab is very different from that in DynaMIT. Since MITSIMLab uses detailed driving behavior models, capacities are not explicitly modeled, but are rather the result of a simulated driver behavior and the geometric characteristics of the network. DynaMIT, however, uses aggregate, mesoscopic speed-density relationships, explicit capacities, and queuing models to represent traffic dynamics. On the demand side, the two systems can employ route choice models that can vary in model structure, explanatory variables, and parameter values. Hence, MITSIMLab provides an unbiased laboratory to test the performance of the proposed calibration methodology.

Typically representative values for various demand and driving behavior parameters were selected as inputs to MITSIMLab in order to generate observed sensor measurements. The time-dependent *historical* profiles of the main OD flow (between nodes O1 and D) and the two side flows were selected so as to generate visible congestion due to the incident, while simultaneously capturing merging and weaving phenomena between the main flow and each of the minor flows. The actual demand levels realized between each OD pair on a particular day were drawn from a distribution with mean flows equal to the historical values. The spread of this distribution, representing the temporal (within-day) variability in demand, was set to 25% around the mean. It was assumed that the corresponding noise was independent across time intervals. Historical travel times on the network were obtained by solving for a stochastic equilibrium based on the historical OD flows (Yang et al., 2000). The actual demand was then simulated together with the other inputs discussed above.

Five link-wide traffic sensors (indicated in Fig. 12.2 by rectangular boxes) provided count and speed observations from MITSIMLab. OD flows in 5-minute time intervals were estimated, yielding 30 flow variables. Data from sensors d1, d2, and d3 were used for this purpose. The performance of the various estimators was analyzed using the Root Mean Square Normalized (RMSN) error statistic across all five sensors:

$$\text{RMSN} = \frac{\sqrt{S \sum_{i=1}^{S} \left(M_i - \mathcal{M}_i \right)^2}}{\sum_{i=1}^{S} M_i} \qquad (12.7)$$

where,

M_i and $\mathcal{M}_i$: elements of the vectors $\left[M_1 \cdots M_h \cdots M_H \right]$ and $\left[\mathcal{M}_1 \cdots \mathcal{M}_h \cdots \mathcal{M}_H \right]$ respectively,

S : the size of either vector.

DynaMIT (Dynamic Network Assignment for the Management of Information to Travelers, Ben-Akiva et al., 2002) was used as the DTA model. OD flows were estimated with the proposed methodology, using sensor count data. If there were incidents during the data collection period, their capacity reduction impacts must be included when calibrating the model. While the drivers were unaware of the incident when they started their trips, the incident reduced the available output capacity of the affected segment. Drivers traveling between O1 and D who chose to travel through the affected link were therefore likely to experience incident delays, depending on their departure times and prevailing traffic conditions. The incident was assumed to remove approximately 55% of the available capacity (this value was estimated in an earlier exercise described by Balakrishna, 2006).

The estimated OD flows compared favorably with the "true" values, reflected in a low RMSN of 0.021, even though the chosen starting flows were far from the optimum. This observation is illustrated in Fig. 12.3 through a visual comparison.

12.4 CASE STUDY II: LOS ANGELES, CALIFORNIA

This section presents a case study on a real network from South Park in downtown Los Angeles, California (Fig. 12.4). The network contains arterials as well as sections of two major interstate highways (the Harbor Freeway, I-110, and the Santa Monica Freeway, I-10). The area is highly congested, due to the regular hosting of sports events and conventions at the Staples Center and the Los Angeles Convention Center.

The network was modeled as a directed graph of 243 nodes and 606 links (divided into 740 segments capturing changes in link cross-section). Most of the urban

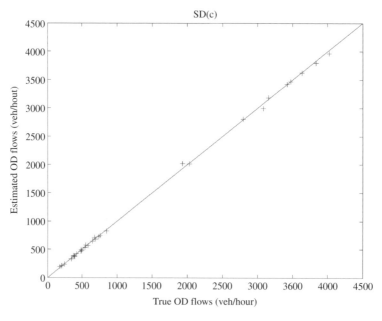

Fig. 12.3 The synthetic network: estimated OD flows.

Fig. 12.4 South park, Los Angeles network [source: www.mapquest.com]

Table 12.1 The fit to counts and speeds (RMSN).

Estimator	Fit to counts		Fit to speeds	
	Freeways	Arterials	Freeways	Arterials
Reference	0.218	0.239	0.181	0.203
Proposed	0.090	0.113	0.088	0.093
% Improv.	58.7	52.7	51.4	54.2

intersections are signalized, and are controlled by the City of Los Angeles through the Automated Traffic Surveillance and Control (ATSAC) system. Time-varying traffic counts and occupancies (densities) for the network were obtained from 203 loop detectors, for September 2004.

The period of interest was chosen as 3:00 AM to 9:00 AM based on analyses of sensor counts by time of day, and includes the AM peak. This period was divided into 15-minute intervals, for a total of $H = 24$ time intervals.

The case study compared the performance of two dynamic OD estimation approaches:

- the reference case, using a sequential, assignment matrix formulation; and

- the proposed methodology, estimating OD flows, route choice and supply parameters simultaneously.

Sensor count data were used in both estimations, and in each case, the fit to count and speed data was analyzed through the RMSN error statistic. The proposed methodology was applied on a Pentium 4 PC with 2 GB of physical memory. A detailed description of the reference case may be found in Gupta (2005).

Table 12.1 summarizes the various RMSN statistics for the two estimators. The results clearly indicate that the proposed methodology outperformed the assignment matrix approach, by improving the fit to counts and speeds by more than 50% uniformly across time and space (on both freeways and arterials).

Apart from improving the fit to counts, the proposed methodology also reduced the error in explaining speed observations. This may be partly attributed to the identification of more accurate OD flows, thereby better replicating the spatial and temporal evolution of densities and traffic dynamics. The accurate modeling of speeds and travel times is critical in the traffic management context, particularly in route guidance applications. The results therefore underscore the various practical benefits of the proposed flexible methodology without assignment matrices. Case studies employing speeds and counts in the objective function of Eq. (12.6) may be found in Balakrishna (2006).

12.5   CONCLUSION

This chapter has presented a systematic OD estimation problem formulation that does not rely on the traditional linear assignment matrix approximation. The complex transformations that map OD flows to traffic measurements were instead captured implicitly through the black-box assignment model itself. This approach allowed the OD estimation procedure to utilize rich data other than link counts, thus improving the accuracy and efficiency of the resulting flow estimates. In addition, the formulation was flexible to accommodate the estimation of other model parameters impacting OD estimation, such as route choice and supply parameters. Computationally efficient algorithms for solving the highly non-linear and large-scale problem were outlined, further permitting the simultaneous estimation of OD flows for several time intervals. The feasibility of the developed methodology was demonstrated through case studies involving the DynaMIT simulation-based traffic assignment model.

12.6   ACKNOWLEDGEMENT

This study was partly funded by the Federal Highway Administration and the Volpe National Transportation Systems Center. The data for the second case study were provided by the Los Angeles Department of Transportation and the PeMS freeway performance measurement system (UC Berkeley and Caltrans, 2005). The authors are responsible for all results and conclusions presented here.

12.7   REFERENCES

Ashok, K. (1996) *Estimation and Prediction of Time-Dependent Origin-Destination Flows, PhD Thesis,* Massachusetts Institute of Technology.

Ashok, K., and Ben-Akiva, M. E. (2000) "Alternative approaches for real-time estimation and prediction of time-dependent origin-destination flows", *Transportation Science,* 34(1):21–36.

Balakrishna, R. (2006) *Off-line Calibration of Dynamic Traffic Assignment Models, PhD Thesis,* Massachusetts Institute of Technology.

Balakrishna, R., Koutsopoulos, H. N., Ben-Akiva, M., Fernandez-Ruiz, B. M., and Mehta, M. (2005) "A simulation-based evaluation of advanced traveler information systems", *Transportation Research Record,* 1910:90–98.

Ben-Akiva, M., Bierlaire, M., Burton, D., Koutsopoulos, H. N., and Mishalani, R. (2001) "Network state estimation and prediction for real-time transportation management applications", *Networks and Spatial Economics,* 1(3/4):291–318.

Ben-Akiva, M., Bierlaire, M., Koutsopoulos, H. N., and Mishalani, R. (2002) "Realtime simulation of traffic demand-supply interactions within DynaMIT". In Gendreau, M. and Marcotte, P., editors, *Transportation and Network Analysis: Miscellenea in honor of Michael Florian*, pp. 19–36. Kluwer, London.

Bierlaire, M., and Crittin, F. (2004) "An efficient algorithm for real-time estimation and prediction of dynamic OD tables". *Operations Research,* 52(1):116–127.

Box, M. J. (1965) "A new method of constrained optimization and a comparison with other methods", *The Computer Journal,* 8(1):42–52.

Cascetta, E. (1984) "Estimation of trip matrices from traffic counts and survey data: a generalized least squares estimator", *Transportation Research,* 18B:289–299.

Cascetta, E., Inaudi, D., and Marquis, G. (1993) "Dynamic estimators of origin-destination matrices using traffic counts", *Transportation Science*, 27(4):363–373.

Cascetta, E., and Nguyen, S. (1988) "A unified framework for estimating or updating origin/destination matrices from traffic counts". *Transportation Research*, 22B(6):437–455.

Cascetta, E., and Postorino, M. N. (2001) "Fixed point approaches to the estimation of O/D matrices using traffic counts on congested networks". *Transportation Science*, 35(2):134–147.

Cascetta, E., and Russo, F. (1997) "Calibrating aggregate travel demand models with traffic counts: estimators and statistical performance", *Transportation*, 24:271–293.

Gupta, A. (2005) *Observability of Origin-Destination Matrices for Dynamic Traffic Assignment, Masters Thesis*, Massachusetts Institute of Technology.

Hazelton, M. L. (2000) "Estimation of origin-destination matrices from link flows on uncongested networks", *Transportation Research*, 34B:549–566.

Hazelton, M. L. (2001) "Inference for origin-destination matrices: estimation, prediction and reconstruction", *Transportation Research*, 35B:667–676.

Huyer, W., and Neumaier, A. (2004) SNOBFIT – stable noisy optimization by branch and fit. *Submitted to ACM Transactions on Mathematical Software.*

Mahmassani, H. S. (2002) "Dynamic network tra±c assignment and simulation methodology for advanced system management applications", Presented at the *81st annual meeting of the Transportation Research Board.*

Park, B. B., Pampati, D. M., Cao, J., and Smith, B. L. (2005) "Field evaluation of DynaMIT in Hampton Roads, Virginia", Presented at the *84th Annual Meeting of the Transportation Research Board.*

Spall, J. C. (1992) "Multivariate stochastic approximation using a simultaneous perturbation gradient approximation", *IEEE Transactions on Automatic Control*, 37(3):332–341.

Toledo, T., Koutsopoulos, H. N., Davol, A., Ben-Akiva, M. E., Burghout, W., Andreasson, I., Johansson, T., and Lundin, C. (2003) "Calibration and validation of microscopic traffic simulation tools: Stockholm case study", *Transportation Research Record*, 1831:65–75.

UC Berkeley and Caltrans (2005) *Freeway Performance Measurement System* (PeMS) 5.4. http://pems.eecs.berkeley.edu/Public, accessed 30th June 2005.

van der Zijpp, N. (1996) *Dynamic Origin-Destination Matrix Estimation on Motorway Networks, PhD Thesis*, Delft University of Technology, Department of Civil Engineering.

van der Zijpp, N. (2002) *Software Package for Estimation of Dynamic OD Matrices,* Delft University of Technology.

Yang, H., Sasaki, T., Iida, Y., and Asakura, Y. (1992) "Estimation of origin-destination matrices from link traffic counts on congested networks", *Transportation Research,* 28B:417–434.

Yang, Q., and Koutsopoulos, H. N. (1996) "A microscopic traffic simulator for evaluation of dynamic traffic assignment systems", *Transportation Research,* 4C(3):113–129.

Yang, Q., Koutsopoulos, H. N., and Ben-Akiva M. E. (2000) "A simulation model for evaluating dynamic traffic management systems", *Transportation Research Record*, 1710:122–130.